Visible Light

Data communications and applications

Series Editor

R Barry Johnson a Senior Research Professor at Alabama A&M University, has been involved for over 50 years in lens design, optical systems design, electro-optical systems engineering, and photonics. He has been a faculty member at three academic institutions engaged in optics education and research, employed by a number of companies, and provided consulting services.

Dr Johnson is an IOP Fellow, SPIE Fellow and Life Member, OSA Fellow, and was the 1987 President of SPIE. He serves on the editorial board of Infrared Physics & Technology and Advances in Optical Technologies. Dr Johnson has been awarded many patents, has published numerous papers and several books and book chapters, and was awarded the 2012 OSA/SPIE Joseph W Goodman Book Writing Award for Lens Design Fundamentals, Second Edition. He is a perennial co-chair of the annual SPIE Current Developments in Lens Design and Optical Engineering Conference.

Foreword

Until the 1960s, the field of optics was primarily concentrated in the classical areas of photography, cameras, binoculars, telescopes, spectrometers, colorimeters, radiometers, etc. In the late 1960s, optics began to blossom with the advent of new types of infrared detectors, liquid crystal displays (LCD), light emitting diodes (LED), charge coupled devices (CCD), lasers, holography, fiber optics, new optical materials, advances in optical and mechanical fabrication, new optical design programs, and many more technologies. With the development of the LED, LCD, CCD and other electo-optical devices, the term ‘photonics’ came into vogue in the 1980s to describe the science of using light in development of new technologies and the performance of a myriad of applications. Today, optics and photonics are truly pervasive throughout society and new technologies are continuing to emerge. The objective of this series is to provide students, researchers, and those who enjoy self-teaching with a wide-ranging collection of books that each focus on a relevant topic in technologies and application of optics and photonics. These books will provide knowledge to prepare the reader to be better able to participate in these exciting areas now and in the future. The title of this series is Emerging Technologies in Optics and Photonics where ‘emerging’ is taken to mean ‘coming into existence,’ ‘coming into maturity,’ and ‘coming into prominence.’ IOP Publishing and I hope that you find this Series of significant value to you and your career.

Visible Light

Data communications and applications

Paul Anthony Haigh
School of Agriculture, Engineering and Mathematics,
Newcastle University, Newcastle-upon-Tyne, United Kingdom

IOP Publishing, Bristol, UK

ISBN 978-0-7503-1680-4 (ebook)
ISBN 978-0-7503-1678-1 (print)
ISBN 978-0-7503-1832-7 (myPrint)
ISBN 978-0-7503-1679-8 (mobi)

DOI 10.1088/978-0-7503-1680-4

Version: 20201201

IOP ebooks

British Library Cataloguing-in-Publication Data: A catalogue record for this book is available from the British Library.

Published by IOP Publishing, wholly owned by The Institute of Physics, London

IOP Publishing, Temple Circus, Temple Way, Bristol, BS1 6HG, UK

US Office: IOP Publishing, Inc., 190 North Independence Mall West, Suite 601, Philadelphia, PA 19106, USA

For David Melber

Contents

Parts of this book were written in the following countries:
London, Newcastle, Manchester, Yeovil, Blandford, United Kingdom
Brussels, Belgium
Sao Paulo, Brazil
Beijing, Shanghai, Qingdao, China
Split, Croatia
Prague, Czech Republic
Paris, France
Cologne, Stuttgart, Germany
Athens, Greece
Hong Kong, Hong Kong
Budapest, Hungary
Dublin, Ireland
Rome, Italy
Vilnius, Lithuania
Amsterdam, Netherlands
Faro, Portugal
Thurwal, Saudi Arabia
Madrid, Barcelona, Spain
Istanbul, Turkey
New York City, San Diego, USA
Hanoi, Vietnam

Preface

Motivated by a rapidly increasing demand for internet connectivity, combined with the hugely over-subscribed radio-frequency (RF) electromagnetic spectrum, researchers have been exploring alternative technologies to support data transmission to end-users, mainly in a home or office environment. One of the most promising candidate technologies to emerge has been visible light communications (VLC), which operates based on the intensity modulation of normal light-emitting diode (LED) luminaries that are becoming commonplace in modern infrastructure. The reason for this candidature is attributed to the fact that the visible spectrum, spanning from ~380 to 760 nm, is around 4000 times larger than that of the RF portion of the spectrum. Furthermore, in contrast to RF, no license is required to operate within the visible portion, and a high degree of frequency re-use is available due to the inherent condition that light cannot pass through opaque objects such as walls.

It is expected that VLC will establish itself as a complementary technology to RF technologies such as Wi-Fi, where multi-technology networks will be deployed in 5th generation (5G) networks, taking advantage of their congruent advantages, such as extremely low latencies, security and Gb/s/user data rates (VLC), together with mobility, non-line-of-sight (NLOS) operation and technology readiness level (Wi-Fi). Therefore it is envisioned that future mobile devices will contain multiple sensors and emitters to take full advantage of such a multi-technology ecosystem.

Remarkable advances have been made in the context of the application of VLC to a telecommunication context. The technology first emerged in the 1990s by demonstration of a small point-to-point link based on a single LED luminary, supporting low data rates. Within the next three decades, transmission speeds have been comfortably raised to gigabits/second (Gb/s) across point-to-multipoint optical wireless access networks, due in no small part to the simultaneous and significant advances in gallium nitride (GaN) technologies, which have enabled LEDs with reasonable bandwidths in the mega-Hertz (MHz) region with high optical power outputs. Due to the relative success of VLC for telecommunications, other applications began to emerge as the technology proliferated, resulting in a broad range of interesting areas, including (but not limited to), localisation and positioning, internet-of-things (IoT), machine-to-machine technologies and healthcare sensing.

Therefore, as these technologies start to broaden, I felt that the time was right to develop a new reference text on the topic that provides coverage on what I feel are the most important topics that will lead to killer applications for VLC. To date, there have been several examples of books published on VLC, all of which aim to cover aspects of telecommunications, without particular or deep focus on new and promising applications. Therefore, this book aims to divide VLC into two distinct and constituent sections: (i) data communications; (ii) applications beyond data communications, which begins the organisation of this book. Next, chapter 2 introduces the technological enablers towards VLC, including advances in GaN

LEDs through the 1990s that enabled the LEDs that are commonly used in VLC today. Chapter 2 will go into detail on the basic theory and operation of such devices, before discussion of new trends for emitter technologies such as multi-drive wavelength-multiplexed single package LEDs, and polymer LEDs (PLEDs) that are becoming popular due to their ultra low costs and the fact they feature several of the advantageous characteristics of plastics and metals simultaneously.

Then begins the first section called *Visible Light for Data Communications*, which explores the necessary building blocks required to create a successful VLC link. Chapter 3 will discuss the necessary circuitry to drive high speed VLC systems, including transmitter design from several viewpoints, following by channel models. The reason that channel models are included at this point, which may seem counterintuitive, is because the characteristics of the LEDs, which were covered in chapter 2, and the channel models, influence both the transmitter and receiver circuits. Therefore, a discussion will be provided on VLC channel models.

In chapter 4, digital signal processing (DSP) will be introduced. DSP is a major technologically relevant area not only for VLC but for telecommunications as a whole and indeed many technologies beyond this, and DSP now supports and advances the vast majority of domains in science as a whole. Hence, a treatment of applicable DSP technologies will be given, and how they influence the signals and data that are received over the VLC links. The evaluation of DSP will begin with an overview of highly spectrally efficient modulation formats commonly used in VLC, adaptive equaliser design and applications of artificial intelligence to links.

One of the main driving forces for VLC as an indoor 5G access network is the fact that LED luminaries offer white light illumination as a bi-product of their data driven emission. Therefore, simultaneous room illumination can be provided alongside data communications. This has divided opinion in the community: some see data communications as the primary function, with room illumination as a secondary function (this is also my opinion), whilst others argue that room illumination is the primary function and data communications is the secondary. LEDs are typically peak-limited and hence, limits are placed on signal-to-noise ratio, peak-to-average power ratio (PAPR) and many other signal characteristics, depending on the lighting conditions This has generated a very interesting area for research focusing on dimming, shadowing and blocking, which will form the basis of the discussion of chapter 5.

VLC can be broadly considered as a broadcasting technology; having lighting infrastructure in the ceiling lends itself very well to supporting a downlink in terms of access networks. This leaves an open question in terms of an uplink. In chapter 6, several of the key technologies that are widely presented as an uplink are discussed. These include RF, infrared or indeed VLC. A thorough discussion of the advantages and disadvantages of each are given and the state-of-the-art is presented. Chapter 6 also concludes the first section of the book, leading to the second section, *Applications Beyond Data Communications*.

The second section will expand the scope of the book beyond telecommunications, looking at emerging applications of VLC including localisation and positioning in chapter 7, which will discuss how VLC can be used for ultra-precision positioning and localisation in passive and active modes. Applications include providing global positioning service (GPS) style directions in indoor and outdoor environments, and surveillance. Subsequently, chapter 8 will cover IoT applications of VLC, which have the potential to solve a number of problems in non-human-interactive scenarios including machine-to-machine and machine-to-infrastructure technologies in order to improve safety in the workplace, amongst other more traditional applications such as data extraction for analytics and optimisation. It will also focus on intelligent transport systems.

This will be followed by chapter 9, which discusses healthcare-based sensors. Developments from every topic previously discussed across the book including materials, structures, devices, circuits and signal processing algorithms have enabled substantial innovation in the healthcare sector, including applications of VLC in medical sensing. This ranges from sensing cerebral palsy in infants through to detection of viral infections based on microfluidics. The state-of-the-art in medical sensing based on VLC will be discussed and evaluated.

Finally, chapter 10 will conclude the book. Many thanks are extended to the authors of papers that are adopted for, or summarised in, this book. Particularly special thanks to my family and friends who have been supportive from the first day I had the chance to write this book.

Author biography

Paul Anthony Haigh

Dr Paul Anthony Haigh is a lecturer in Communications at the Intelligent Sensing and Communications (ISC) group at Newcastle University since March 2019.

He is an Associate Editor for *Frontiers in Optical Communications & Networks*, IET *Electronics Letters* and a section Editor for MDPI *Sensors*. He has published more than 100 articles in respected periodicals such as Nature Publishing's *Light: Science & Applications*, *IEEE Journal on Selected Areas in Communications*, *IEEE Wireless Communications* and *IEEE Communications Magazine*. Paul is an early-career academic with multi-disciplinary skills and has contributed significantly to the field of optical communications over the past decade, demonstrating a strong track record of academic excellence. His research has focused on highly spectrally efficient communications systems with particular focus on visible light communication with applied digital signal processing and artificial intelligence.

Before he moved to Newcastle, he was a Research Fellow at University College London from 2016 to 2019 where he worked on the £1.25M EPSRC funded project 'MARVEL', which focused on developing technologies to enable multi-functional polymer light-emitting diodes that are capable of high rate communication and active lighting. Prior to this, he was a Senior Research Associate at the University of Bristol from 2014 to 2016, where he was named as a researcher co-investigator on the £1.6M EPSRC funded project 'INITIATE', to which he contributed to the case for support and designed the project's technical approaches.

From 2011 to 2014, Paul undertook his PhD research at Northumbria University where he developed, along with colleagues from University College London and Siemens Healthineers, a new area in VLC that makes use of new organic polymer devices. The challenge uncovered was that PLEDs are restricted to several kHz of modulation bandwidth and hence, achieving high speed links was not trivial. Despite this, Paul was able to drive achievable bit rates into the high Mb/s region through intelligent application of electronic circuit, signal processing and artificial intelligence techniques.

Finally, Paul started his research career (2010–2011) at the European Organisation for Nuclear Research (CERN), where he undertook the prestigious Marie Curie Fellowship. In this period, he performed research on ultra-high-speed optical links in an extremely radiative environment, contributing to the design of radiation-hard optoelectronic circuits for the large hadron collider (LHC) experiments.

List of acronyms

Acronym	Description
μLED	Micro-light-emitting diode
m-CAP	Multi-band carrier-less amplitude and phase modulation
5G	Fifth generation
ac	Alternating current
ACG	Automatic gain control
ADC	Analogue-to-digital converter
ANN	Artificial neural network
AOA	Angle of arrival
APD	Avalanche photodiode
AWGN	Additive white Gaussian noise
BER	Bit-error rate
BHJ	Bulk heterojunction
BLW	Baseline wander
BP	Back-propagation
BPF	Band-pass filter
CAGR	Compound annual growth rate
CAP	Carrier-less amplitude and phase modulation
CCR	Continuous current reduction
CDF	Cumulative distribution function
Ce:YAG	Cerium-doped yttrium aluminium garnet
CIE	Commission Internationale de l'Éclairage
CP	Cyclic prefix
CPAM	Coloured pulse amplitude modulation
CPE	Common phase error
CSK	Colour shift keying
DAC	Digital-to-analogue converter
dc	Direct current
DOS	Density of states
DSP	Digital signal processing
EM	Electromagnetic
EPSRC	Engineering & Physical Sciences Research Council
EVM	Error vector magnitude
FBMC	Filter-bank multi-carrier
FDM	Frequency division multiplexing
FEC	Forward error correction
FFT	Fast Fourier transform
FIR	Finite impulse response
FOFDM	Fast orthogonal frequency division multiplexing
FoV	Field-of-view
FPGA	Field programmable gate array
FSO	Free-space optics
FTN	Faster-than-Nyquist
GaAs	Gallium arsenide
GaN	Gallium nitride
GaP	Gallium phosphide
Ge	Germanium

GNSS	Global navigation satellite system
GPS	Global positioning system
HOMO	Highest occupied molecular orbital
HPF	High-pass filter
IBI	Inter-band interference
ICI	Inter-carrier interference
IDCT	Inverse discrete cosine transform
IEEE	Institute of Electronic & Electrical Engineers
IFFT	Inverse fast Fourier transform
IM	Intensity modulation
InGaAs	Indium gallium arsenide
InGaN	Indium gallium nitride
IoT	Internet-of-things
IR	Infrared
ISI	Inter-symbol interference
ISO	International Organisation for Standardisation
ITO	Indium tin oxide
ITS	Intelligent transport system
LCAO	Linear combination of atomic orbitals
LD	Laser diode
LED	Light-emitting diode
LiRa	Light-radio
LM	Levenberg–Marquardt
LMS	Least mean squares
LOS	Line-of-sight
LPF	Low-pass filter
LUMO	Lowest unoccupied molecular orbital
M2M	Machine-to-machine
MAC	Medium access protocol
MIMO	Multiple-input multiple-output
MLP	Multi-layer perceptron
MRC	Maximal-ratio-combining
MSE	Mean squared error
NEP	Noise equivalent power
NIC	Negative impedance converter
NIR	Near infrared
NLOS	Non-line-of-sight
OFDM	Orthogonal frequency division multiplexing
OLED	Organic light-emitting diode
OOK	On-off keying
OPD	Organic photodetector
OPV	Organic photovoltaic
OQAM	Offset-quadrature amplitude modulation
OSI	Open systems interconnection
P3HT	Poly(3-hexylthiophene-2,5-diyl)
PAM	Pulse amplitude modulation
PAPR	Peak-to-average power ratio
patl	Pseudo-artificial transmission line
PCBM	[6,6]-phenyl-C61-butyric acid methyl ester
PD	Photodetector

PDF	Probability density function
PIN	Positive-intrinsic-negative
PLED	Polymer light-emitting diode
PPM	Pulse position modulation
PRBS	Pseudo-random binary sequence
PV	Photovoltaic
PWM	Pulse width modulation
QAM	Quadrature amplitude modulation
QoS	Quality of service
RAGB	Red-amber-green-blue
ReLu	Rectified linear unit
RF	Radio frequency
RGB	Red-green-blue
RLC	Resistor-inductor-capacitor
RLS	Recursive least squares
RMS	Root mean square
RRC	Root-raised cosine
RSS	Received signal strength
RSSI	Received signal strength indicator
s-CAP	Staggered carrier-less amplitude and phase modulation
SCG	Scaled conjugate gradient
SDR	Software-defined radio
SEFDM	Spectrally efficient frequency division multiplexing
Si	Silicon
SIM	Subcarrier index modulation
SINR	Signal-to-interference-plus-noise ratio
SN-CAP	Super-Nyquist carrier-less amplitude and phase
SNR	Signal-to-noise ratio
TCP	Transmission control protocol
TDMA	Time-division multiple access
TDOA	Time difference of arrival
TIA	Transimpedance amplifier
UDP	User datagram protocol
UK	United Kingdom
V2I	Vehicle-to-infrastructure
V2V	Vehicle-to-vehicle
V2X	Vehicle-to-X
VLC	Visible light communications
VLCC	Visible light communications consortium
VPPM	Variable pulse position modulation
WDM	Wavelength division multiplexing
Wi-Fi	Wireless fidelity
WPLED	White-phosphor light-emitting diode
ZFE	Zero-forcing equaliser

Chapter 1

Introduction

1.1 Introduction to visible light communications

In recent years, visible light communications (VLC) has emerged as an excellent candidate to complement traditional radio frequency (RF) wireless access networks and as a result has become the focus of enormous research attention [1–6]. RF technologies such as wireless fidelity (Wi-Fi) are well established and have been widely adopted worldwide in both home and office environments. As a result, Wi-Fi is commercially available at a low cost and offers several advantages such as relatively high speeds in the region of hundreds of Mb/s, wide coverage supporting mobility and ease-of-use. The main challenge with RF technologies is rooted in their limited spectrum availabilities, particularly with 5th generation (5G) new radio adding a further layer of complication to already crowded resources. Cisco perform an annual survey of mobile data traffic, using it to forecast the quantity of mobile data expected to be transmitted per-month over the coming years. For instance, the 2019 edition forecasts that mobile data demand will exceed 75 EB/month by 2022 [7]. The report also shows that demand is increasing exponentially with a compound annual growth rate (CAGR) of 46%, and this analysis is illustrated here in figure 1.1. This is not a temporary trend and will not be easily ameliorated either, as an ever-increasing number of mobile devices require internet connectivity and emerging technologies such as the internet-of-things (IoT) are only increasing demand. This is compounded by the fact that the RF spectrum is limited from 0.3 to 3 GHz (or 0.3 to 300 GHz including microwave radiation), and the vast majority of the spectrum has already been allocated. Figure 1.2 is adopted from Roke Manor Research [8], which highlights the allocation of the spectrum and the serious overcrowding within RF, leading to extremely high license fee premiums.

1.1.1 Solutions to the spectrum crunch

Researchers commonly accept that two solutions emerge; firstly, searching for alternative spectra, such as optical wavelengths, or secondly, increasing the spectral

doi:10.1088/978-0-7503-1680-4ch1

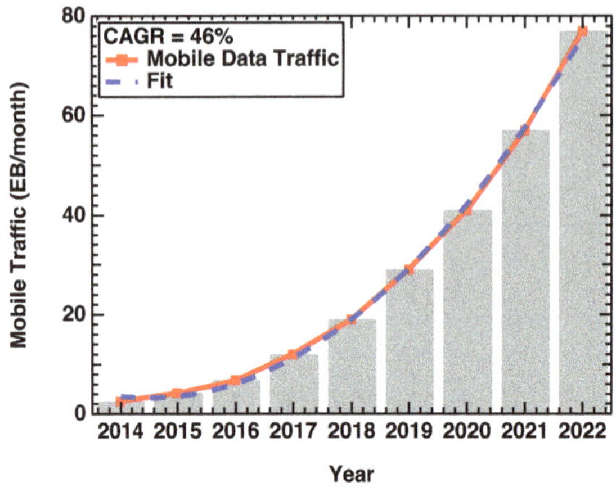

Figure 1.1. Predicted traffic growth until 2022, and the vast majority of this traffic is expected to originate or terminate in an indoor environment. [7] Reproduced with permission © Springer.

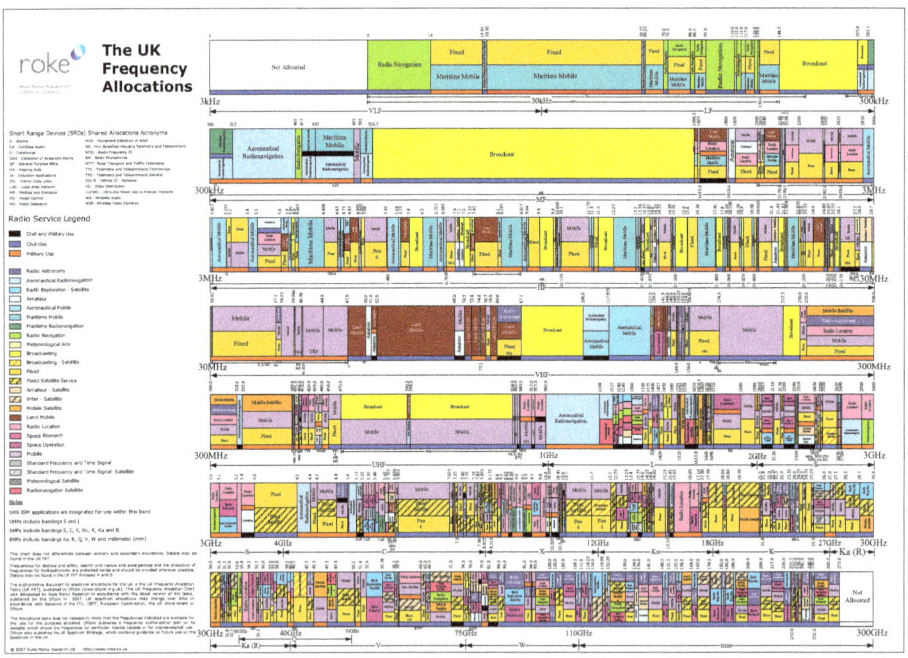

Figure 1.2. The UK RF and microwave allocation [8], showing a clear requirement for additional spectrum due to overcrowding. If supply is to cope with demand additional spectrum will be required, or improved spectral efficiencies should be delivered.

efficiency of current modulation formats in order to better use the available spectrum. The latter has been widely explored in the literature, with the wide adoption of highly spectrally efficient modulation formats such as orthogonal frequency division multiplexing (OFDM) [9–11], which, when used with adaptive

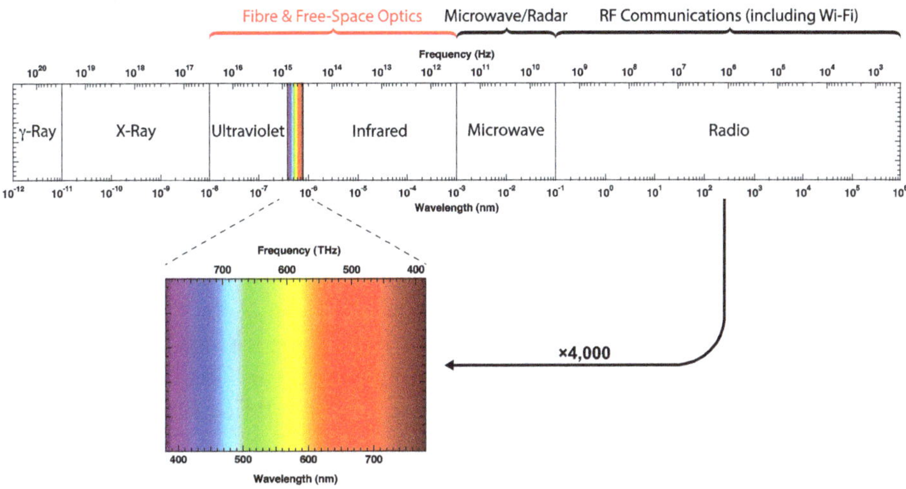

Figure 1.3. The electromagnetic spectrum with wavelengths and frequencies assigned to different technological domains, the visible spectrum spans approximately 400 Thz, around 4000 times higher than that of the RF and microwave range.

bit- and power-loading techniques [12] can offer spectral efficiency gains over pulse-based modulation formats. Moreover, researchers have gone a step further, introducing non-orthogonal modulation formats such as faster-than-Nyquist (FTN) [13–15] and spectrally efficient frequency division multiplexing (SEFDM) [16, 17], which purposely violate the orthogonality conditions, introducing controlled self-interference. The former method, searching for alternative spectra to utilise, is the most practical and there are several options available in the electromagnetic (EM) spectrum, illustrated in figure 1.3. The most realistic of these are the optical domain wavelengths, broadly classified into ultraviolet, visible and infrared (IR). The ultraviolet can be largely dismissed for commercial telecommunications due to strict eye safety limits [18] and high power requirements, although it does have military and long range applications due to backscattering of light, which is not covered here but can be referred to in [19]. At the same time, IR technologies are extremely popular for telecommunications and are widely adopted in optical fibre networks for transporting large quantities of data at high speeds over long distances [20, 21].

These are normally concentrated into three separate bands; 850 nm, 1310 nm and 1550 nm. Due to the lower power loss in optical fibres, the two higher wavelengths are generally used in optical fibre systems [22], while 850 nm is typically reserved for free-space operation [23, 24]. One of the reasons that IR devices are popular is that devices are commercially available that offer wider bandwidths than both RF and visible wavelengths, in the order of GHz. It is restricted by eye safety limits similar to ultraviolet, however, and generally the sources are laser diodes (LDs), which are point sources, and require additional optics such as lens and diffusers to control the beams and produce reasonable beam patterns for indoor applications. Such additional optics are bulky, and generally bespoke, making widespread adoption difficult without tailoring the system to its environment. Hence, after the conception of indoor IR access by Gfeller and Bapst in 1979 [23], its subsequent popularisation in 1997 by Kahn and Barry [24], and standardisation [25], alternative spectrum was still sought and indoor IR never received the widespread adoption its potential perhaps warranted.

1.1.2 Emergence of visible light communications

A few years later in 2000, VLC was proposed in Japan by Tanaka, Haruyama and Nakagawa in [26] as a proposition to use the visible range of the EM spectrum, i.e. using the 380–780 nm range of wavelengths for data transmission. This corresponds to a bandwidth exceeding 300 THz, which is unregulated and license free, meaning there are no fees associated with using the spectrum as in RF, which is a significant advantage.

A brief timeline and the technological enablers leading to and subsequent progress in VLC is outlined in figure 1.4 starting with the invention of the photophone by Alexander Graham Bell in [27], which modulated sound onto an optical carrier (in this case, the Sun) via vibration of a material for transmission over free space. The telephone, also invented by Bell, turned out to be a far more popular invention and optical communications remained only a concept until, between 1958 and 1960, the laser was introduced approximately simultaneously by four different groups [28–31]. This led to the popularisation of free-space optics (FSO), another name for outdoor optical communication using IR wavelengths, in the 1970s, which is still an active research topic today. As has been described, the next steps were provided by Gfeller and Bapst [23] and then Kahn and Barry [24] over the following decades followed by the invention of VLC by Tanaka, Haruyama and Nakagawa [26]. In 2003, the visible light communications consortium (VLCC) was proposed with an aim to gather major players from academia and industry with a view to standardisation. Developments were ongoing until 2008, when the first major European grant involving VLC was announced, the OMEGA project, which aimed to provide optical wireless access to the indoor environment based on near infrared (NIR) and VLC. In 2009, the Institute of Electronic and Electrical Engineers (IEEE) called for standardisation of VLC through IEEE 802.15.7 [32], which was

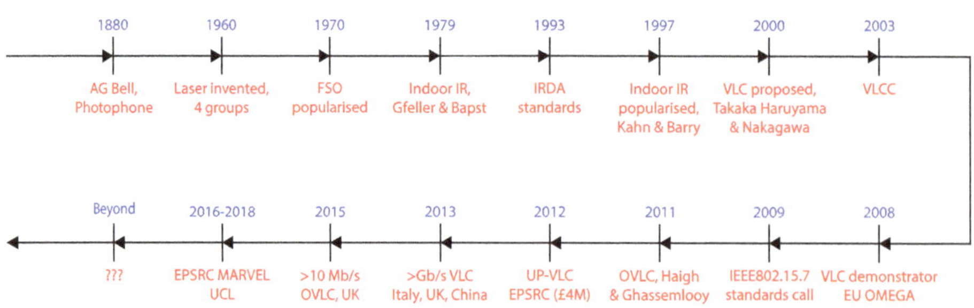

Figure 1.4. A (somewhat incomplete) timeline of technological advances that have led to the current state-of-the-art in VLC systems. The most important of which was the general inception of VLC by Tanaka *et al* in 2000.

subsequently submitted in 2011. In the same year, a new concept within VLC was introduced by the author who first proposed to use organic light-emitting diodes (OLEDs) as the transmitter instead of conventional inorganic light-emitting diodes (LEDs) based on advantages such as mechanical flexibility and arbitrary photo-active surface areas [33]. The disadvantage of OLEDs in VLC is significantly reduced bandwidths, as will be discussed later. Next, in 2012, a United Kingdom (UK) programme grant funded by the Engineering and Physical Sciences Research Council (EPSRC) provided ~£5M to explore and develop micro-LEDs (μLEDs) for use in VLC systems, eventually resulting in a 3 Gb/s link, which was a world record at the time [34]. Subsequently, globally, Gb/s links were demonstrated worldwide from key VLC groups in the UK, Italy and China, with several, non-exhaustive example references included as follows [34–36]. Later, in 2015, the author was able to demonstrate relatively high transmission speeds using OLEDs based on polymers, known as polymer light-emitting diodes (PLEDs), which resulted in a ~55 Mb/s data rate [37]. This landmark led to the second major VLC grant within the UK, which proposes to revolutionise systems based on PLEDs and obtained £1.25M from the EPSRC. Beyond this in the next few years, there is huge potential for the expansion of VLC into numerous domains, some of which will be covered in this book. The research community is rapidly expanding and beginning to find new applications in telecommunications and beyond.

A block diagram for a generic, simplified VLC system is illustrated in figure 1.5. The data source to be transmitted must first be compressed by a source code to reduce the necessary transmission bandwidth and then protected against channel errors induced by noise and interference by a channel code. Neither source nor channel codes are covered in this book; however, there are well established texts in the literature that discuss modern codes in detail [14, 38], while the subject was initiated by Shannon in his seminal text, '*The Mathematical Theory of Communication*' [39]. After coding is performed, the data must be encoded by a

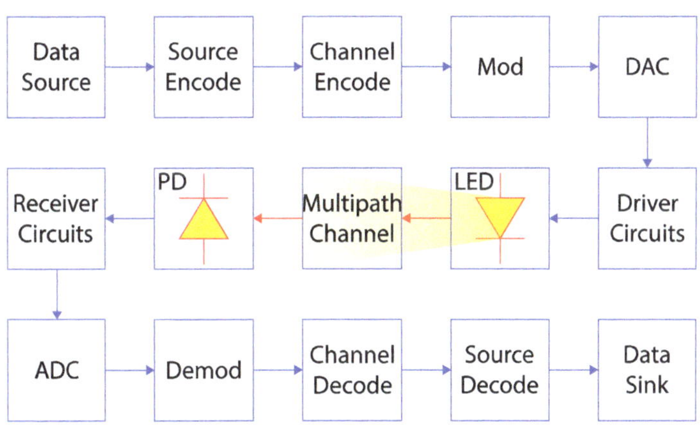

Figure 1.5. A generic block diagram of a VLC system that includes all building blocks required for high quality information transmission. Acronyms/abbreviations: Mod (modulation), DAC (digital-to-analogue converter), ADC (analogue-to-digital converter), Demod (demodulation).

modulation format (Mod in figure 1.5) and is subsequently converted from the digital domain into the analogue domain via a digital-to-analogue converter (DAC). The analogue data is then fed into the driving circuit of the LEDs, which typically consists of active and passive circuit elements that amplify the alternating current (ac) signal as desired according to the individual characteristics of the emitters, followed by a direct current (dc) bias, since optics must be non-negative, as there can be no negative optical signal. The dc-biased ac signal is then transmitted over the channel, which is modelled as multi-path, additive white Gaussian noise (AWGN) and a flat-fading channel for the frequencies generally considered in VLC.

The receiver in VLC systems is called a photodetector (PD) and they generally consist of silicon (Si), which is a photo-active material that absorbs radiation across the entire visible and NIR wavelength range up to ~1100 nm. The photocurrent generated by the PD is then translated into a voltage by a transimpedance amplifier (TIA) and then out-of-band is generally eliminated by a low-pass filter (LPF), which makes up the receiver circuits. Afterwards, the signal is translated into the digital domain by a analogue-to-digital converter (ADC) before demodulation, error detection and correction, source de-compression and finally the data used via the medium it was intended, i.e. audio or video, to name two examples.

1.2 State-of-the-art in visible light communications

VLC has been driven by advances in LED technologies such as gallium nitride (GaN), indium gallium nitride (InGaN) and gallium phosphide (GaP). As these technologies matured steadily since the 1970s, more highly efficient devices were produced until relatively high optical power could be produced at low costs. This is reflected in figure 1.6, which shows the gradual and consistent improvement of LED luminous efficacy over several decades for several materials. Values of luminous efficacy are now far in excess of those found in the best fluorescent and incandescent bulbs and their lifetimes are also substantially longer.

Furthermore, in figure 1.7 it is clear that the amount of luminous flux that can be extracted from a single luminary has increased approximately twentyfold every decade for a number of different commercially available technologies. In conjunction with this, as illustrated in figure 1.8 due to increasing mass production, the luminous flux per dollar has also increased at half this rate, approximately tenfold per decade over the last 50 years.

Leveraging on these advances, in order to make use of the extra bandwidth provided by the visible range of the EM spectrum, VLC typically makes use of commercially available LEDs to convert electronic signals into optical intensity, in a process known as intensity modulation (IM). In the original paper [26], low power LEDs were used to demonstrate a proof-of-concept; however, these are not the main sources used in VLC systems. There are two far more common options, both of which offer high optical power outputs suitable for illumination of indoor-scale environments: (i) GaN-based LEDs, which underwent significant development in the 1990s [41], can offer wide bandwidths in the mid-MHz region. These are the most common emitter devices used in VLC systems due to their cost effectiveness and simplicity of

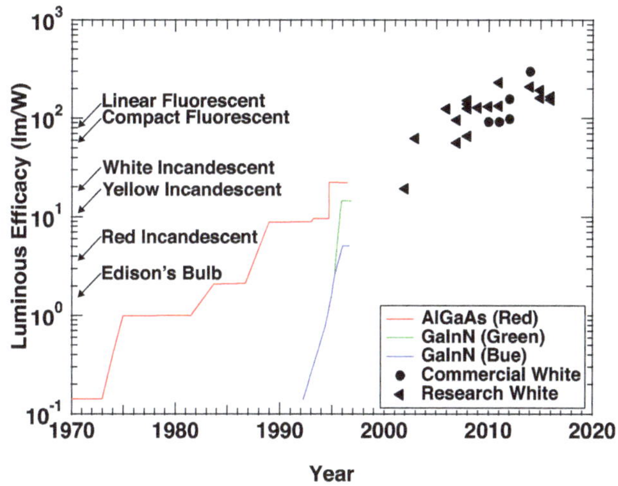

Figure 1.6. Progress of LED technologies with time as a function of luminous efficacy. There has clearly been substantial progress with achieving high efficacies in both commercial and academic settings. Adapted from [40].

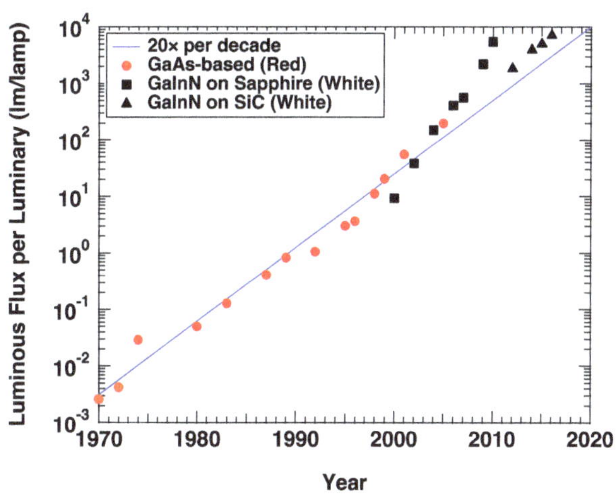

Figure 1.7. Different LED semiconductor technology luminous flux production as a function of time. Adapted from [40].

use. On the other hand, in order to obtain white-light emission, which is useful for room illumination, the GaN emissive layer must be coated with a colour-converting phosphor, typically cerium-doped yttrium aluminium garnet (Ce:YAG) [42], which absorbs some of the blue wavelengths and re-emits at longer yellowish wavelengths and are commonly known as white-phosphor light-emitting diodes (WPLEDs). The aggregate of this is white-light emission, as is highlighted in figure 1.9. The other common emitter packages are (ii), red-green-blue (RGB) or red-amber-green-blue (RAGB)-LEDs, also shown in figure 1.9. These packages are slightly different from (i) because they contain several individually and independently addressable LEDs that

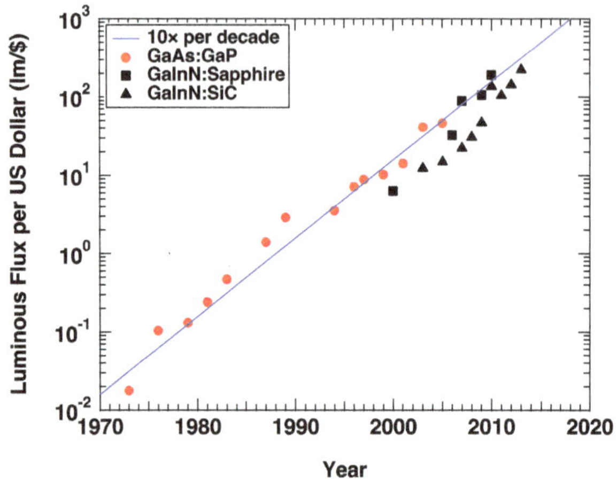

Figure 1.8. Clearly, not only have LEDs got brighter with time, they have also become cheaper at a slightly lower rate. Adapted from [40].

Figure 1.9. Two popular methods to produce white light; firstly the white phosphor approach where a blue wavelength chip is coated with a yellowish colour-converting phosphor, and the RAGB approach where a single chip contains four different devices that provide white light through summed illumination.

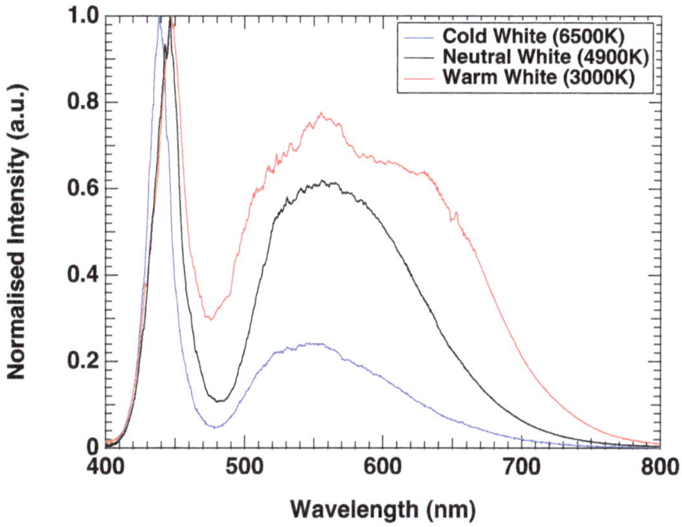

Figure 1.10. Optical spectral response of three different temperate white-phosphor devices with an increasingly thick phosphor layer.

require bespoke driving circuits matched to the characteristics of each chip. While this adds circuit complexity, there are two significant advantages, which are as follows. Firstly, each chip tends to have relatively wide bandwidth in the mid-MHz region and is not perturbed by a slow phosphor conversion process, which means that high data rates are more easily achieved considering that independent data can be transmitted on each wavelength [35]. The second advantage is that by adjusting the intensity of each coloured chip appropriately, colour tuning can be achieved to vary the temperature of the light as desired, which can have a considerable effect on mental health and wellbeing in humans [43]. Moreover, Popoola [44] showed that variation in colour temperature has little impact on the quality of the communications link and hence, colour tuning can be set as desired by the user.

Research has focused on both types of emitter, WPLEDs and RAGB-LEDs and a short discussion of both will be given, with further details later in chapter 2. The most popular emitter in VLC are WPLEDs because of their cost effectiveness and simplicity, as mentioned. An example of the optical intensity spectrum is shown in figure 1.10 and the two component parts are clear; a blue peak at 450 nm and the white-phosphor emission peaking around 550 nm. They also offer high optical power but the trade-off is a low bandwidth in the low MHz region, as mentioned, due to the white-phosphor conversion, which is a slow process. Research has been developed that investigates removing the slow yellowish phosphor component of the signal by isolating the blue part using a dichroic filter with remarkable success, enabling data rates >100 Mb/s for the first time [45]. The dichroic filter is generally known colloquially as a 'blue filter'. In the literature, this technique was widely adopted in early VLC literature [45–50]. In [45] in particular, the blue filtering technique was combined with a passive equalisation method to improve the system bandwidth by reducing the effective plate capacitance of the WPLEDs whilst isolating only the fast blue temporal component of the signal. The report in [45] prompted an explosion into research both into the use of blue filtering and equalisation techniques [33, 36, 45–54], of which there are many and they will be introduced later in chapter 4. Since the first demonstration of 100 Mb/s in [45], gigabit transmission speeds using WPLEDs have been also been reported based on novel driving circuits, equalisers, blue filters and advanced modulation formats. For instance, in [36, 48], both report present Gb/s links based on WPLEDs and utilise advanced modulation formats.

In the initial demonstration of blue filtering and equalisation [45], on-off keying (OOK) was used, which is the most simple modulation format and consists of a pulse of energy over a bit period to signify the transmission of a logic-1 and the absence of a pulse to signify the transmission of a logic-0. This will be covered later in detail in chapter 4. One of the main problems with OOK is that it offers a single bit-per-symbol and therefore potentially doesn't utilise the bandwidth as efficiently as possible, if sufficient signal-to-noise ratio (SNR) is available. Thus, [36] improves this by making use of OFDM, a system that transmits information in parallel over a number of subcarriers, where each subcarrier carries independent information. The subcarriers are separated by integer multiples of $1/T$, where T is the symbol period. This method enables potentially higher data rates as each subcarrier can carry a different number of bits-per-symbol based on the channel state information over the frequency in which it exists. Using OFDM in conjunction with analogue equalisation circuit design, the method in [36] was able to transmit a data rate of 1.6 Gb/s. Furthermore, an alternative advanced modulation format gaining increasing interest in recent years is carrier-less amplitude and phase modulation (CAP), which makes use of a Hilbert pair to transmit two signals (traditionally) in the same frequency range but separated in phase by 90° as shown in [48], which shows a transmission speed of 1.1 Gb/s. Incidentally, the fact that the data rate reported for OFDM is in excess of that over CAP does not indicate superiority of the modulation format as external factors must be considered, such as the system bandwidth, transmission distance and optical power.

In recent years, alternative colour conversion techniques have been proposed including using polymer-based converters [55–57]. Research has also been presented in a number of reports that argues whether blue filtering is the best method for obtaining high data rates in systems utilising WPLEDs [51, 52]. The reason for this is that advanced modulation formats such as OFDM or advanced equalisation techniques actually make use of the high SNR that is obtained by retaining the entire transmitted optical spectrum. Considering that dichroic filtering removes up to 50% of the signal power [58], most of the SNR is sacrificed to obtain a higher bandwidth. Formats like OFDM can utilise the higher SNR by loading a higher number of bits-per-symbol-per-subcarrier, regardless of the lower bandwidth, and as such, demonstrate higher transmission speeds in comparison to an equivalent link with blue filtering [51]. Furthermore, equalisation techniques that perform training based on received data can also make use of the higher SNR to demonstrate higher transmission speeds, as shown in [52]. Therefore, there are considerable design considerations to take into account when using WPLEDs in a VLC system.

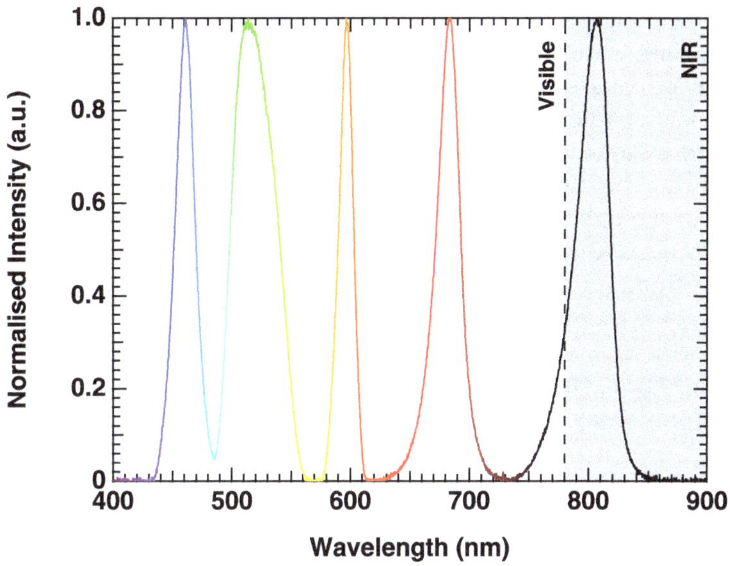

Figure 1.11. The optical spectrum of an RAGB device with a near-infrared component included for illustrative purposes.

Figure 1.11 shows an example of a highly isolated (i.e. low cross-talk) RAGB optical spectra with a NIR wavelength added to highlight the potential of combining technologies for increased data rate. When considering RAGB-LEDs, one may omit considerations of dichroic filters, as they are not required to obtain white light. High data rates have also been demonstrated using both RGB- and RAGB-LEDs in the literature, using the same modulation formats (i.e. OFDM and CAP) previously outlined [2, 35, 47, 59–61] and hence the details will not be covered. Generally, higher data rates can be achieved (<10 Gb/s to date [62]) using RAGB-LEDs due to the additional bandwidth offered by each wavelength and the parallel transmission of data on each isolated wavelength. One of the key issues in RAGB VLC systems is cross-talk, which is where the energy in each wavelength overlaps, causing superposition of the independent data that is ideally transmitted in isolation. The impact of cross-talk is discussed in detail in [63], which provides mathematical models for the emission spectra of multi-wavelength LEDs and an expression for cross-talk based on these. The radiation emitted by the LEDs is transmitted over a free-space channel, which is typically indoors over a few metres. Next, an introductory discussion on the channel will be provided.

1.3 Summary

For this introductory chapter, a brief overview of the background of VLC including the context in which it sits and the state-of-the-art of what has been reported in the research community has been given. The underpinnings of most of these systems are provided in later chapters. In the next chapter, the physical mechanisms that ensure the operation are discussed including light generation and absorption by the photo-active components. In the succeeding chapters, high speed circuit design and channel modelling are considered before advanced modulation formats are discussed. Some space is devoted to equalisation techniques and machine learning before additional chapters on acquiring a good balance between lighting and data communication systems. This is fundamentally important because the advantages of VLC are mostly advertised as a dual purpose, dual benefit system, and therefore optimisation of the light that the users receive is extremely important. Afterwards, chapter 6 discusses the candidate uplink technologies and the state-of-the-art developments from researchers, globally. The final chapters of the book are on localisation, where extremely high resolution indoor mapping can be performed using active or passive mode systems, then intelligent transport systems for safety and inter-vehicle transmission are discussed. The book is concluded with a chapter on oxygen sensing using visible light.

References

[1] Tsiatmas A, Baggen C P M, Willems F M J, Linnartz J P M G and Bergmans J W M 2014 An illumination perspective on visible light communications *IEEE Commun. Mag.* **52** 64–71

[2] Karunatilaka D, Zafar F, Kalavally V and Parthiban R 2015 LED based indoor visible light communications: state of the art *IEEE Commun. Surv. Tutor.* **17** 1649–78

[3] Jovicic A, Li J and Richardson T 2013 Visible light communication: opportunities, challenges and the path to market *IEEE Commun. Mag.* **51** 26–32

[4] Grobe L, Paraskevopoulos A, Hilt J, Schulz D, Lassak F, Hartlieb F, Kottke C, Jungnickel V and Langer K D 2013 High-speed visible light communication systems *IEEE Commun. Mag.* **51** 60–6

[5] Armstrong J, Sekercioglu Y A and Neild A 2013 Visible light positioning: a roadmap for international standardization *IEEE Commun. Mag.* **51** 68–73

[6] Wu S, Wang H and Youn C H 2014 Visible light communications for 5G wireless networking systems: from fixed to mobile communications *IEEE Netw.* **28** 41–5

[7] Wu S, Wang H and Youn C H 2017 Cisco visual network index: global mobile data traffic forecast, 2016–2021 *Cisco Public Information White Paper*

[8] 2017 *UK Radio Frequency Allocations Chart* (Gloucester: Roke Manor Research Ltd.)

[9] Mossaad M S A, Hranilovic S and Lampe L 2015 Visible light communications using OFDM and multiple LEDs *IEEE Trans. Commun.* **63** 4304–13

[10] Hong Y, Wu T and Chen L K 2016 On the performance of adaptive MIMO-OFDM indoor visible light communications *IEEE Photonics Technol. Lett.* **28** 907–10

[11] Huang X, Shi J, Li J, Wang Y, Wang Y and Chi N 2015 750 Mbit/s visible light communications employing 64QAM-OFDM based on amplitude equalization circuit *Optical Fiber Communication Conf.* (Optical Society of America) p Tu2G.1

[12] Bykhovsky D and Arnon S 2014 An experimental comparison of different bit-and-power-allocation algorithms for DCO-OFDM *J. Light. Technol.* **32** 1559–64

[13] Mazo J E 1975 Faster-than-Nyquist signaling *Bell Syst. Tech. J.* **54** 1451–62

[14] Anderson J B, Rusek F and Öwall V 2013 Faster-than-Nyquist signaling *Proc. IEEE* **101** 1817–30

[15] Zhou J, Qiao Y, Yang Z and Sun E 2016 Faster-than-Nyquist non-orthogonal frequency-division multiplexing based on fractional Hartley transform *Opt. Lett.* **41** 4488–91

[16] Rodrigues M and Darwazeh I 2003 A spectrally efficient frequency division multiplexing based communications system *Proc. 8th Int. OFDM Workshop*

[17] Wang Y, Zhou Y, Gui T, Zhong K, Zhou X, Wang L, Lau A P T, Lu C and Chi N 2016 Efficient MMSE-SQRD-based MIMO decoder for SEFDM-based 2.4-Gb/s-spectrum-compressed WDM VLC system *IEEE Photonics J.* **8** 1–9

[18] Sliney D H and Mellerio J 2013 *Safety with Lasers and other Optical Sources: A Comprehensive Handbook* (Berlin: Springer)

[19] Xu Z and Sadler B M 2008 Ultraviolet communications: potential and state-of-the-art *IEEE Commun. Mag.* **46** 67–73

[20] Liu Z, Erkilinc M S, Kelly B, O'Carroll J, Phelan R, Thomsen B C, Killey R I, Richardson D J, Bayvel P and Slavik R 2016 49 Gbit/s direct-modulation and direct-detection transmission over 80 km SMF-28 without optical amplification or filtering *ECOC 2016; 42nd European Conf. on Optical Communication* pp 1–3

[21] Liu Z *et al* 2017 Record high capacity (6.8 Tbit/s) WDM coherent transmission in hollow-core antiresonant fiber *2017 Optical Fiber Communications Conf. and Exhibition (OFC)* pp 1–3

[22] Agrawal G P 2012 *Fiber-Optic Communication Systems* vol 222 (New York: Wiley)

[23] Gfeller F R and Bapst U 1979 Wireless in-house data communication via diffuse infrared radiation *Proc. IEEE* **67** 1474–86

[24] Kahn J M and Barry J R 1997 Wireless infrared communications *Proc. IEEE* **85** 265–98

[25] Williams S 2000 IrDA: past, present and future *IEEE Pers. Commun.* **7** 11–9

[26] Tanaka Y, Haruyama S and Nakagawa M 2000 Wireless optical transmissions with white colored LED for wireless home links *11th IEEE International Symp. on Personal Indoor and Mobile Radio Communications. PIMRC 2000. Proc. (Cat. No.00TH8525)* vol 2 pp 1325–9

[27] Graham Bell A 1880 The photophone *J. Franklin Inst.* **110** 237–48

[28] Schawlow A L and Townes C H 1958 Infrared and optical masers *Phys. Rev.* **112** 1940–9

[29] Sanders J H 1959 Optical maser design *Phys. Rev. Lett.* **3** 86–7

[30] Javan A 1959 Possibility of production of negative temperature in gas discharges *Phys. Rev. Lett.* **3** 87–9

[31] Maiman T H 1960 Optical and microwave-optical experiments in ruby *Phys. Rev. Lett.* **4** 564–6

[32] IEEE Standard 2011 IEEE standard for local and metropolitan area networks–part 15.7: Short-range wireless optical communication using visible light *IEEE Std 802.15.7-2011* pp 1–309

[33] Haigh P A, Ghassemlooy Z, Le Minh H, Rajbhandari S, Arca F, Tedde S F, Hayden O and Papakonstantinou I 2012 Exploiting equalization techniques for improving data rates in organic optoelectronic devices for visible light communications *J. Lightwave Technol.* **30** 3081–8

[34] Tsonev D *et al* 2014 A 3-Gb/s single-LED OFDM-based wireless VLC link using a gallium nitride μLED *IEEE Photonics Technol. Lett.* **26** 637–40

[35] Cossu G, Khalid A M, Choudhury P, Corsini R and Ciaramella E 2012 3.4 Gbit/s visible optical wireless transmission based on RGB LED *Opt. Express* **20** B501–6

[36] Huang X, Wang Z, Shi J, Wang Y and Chi N 2015 1.6 Gbit/s phosphorescent white LED based VLC transmission using a cascaded pre-equalization circuit and a differential outputs PIN receiver *Opt. Express* **23** 22034–42

[37] Haigh P A, Bausi F, Le Minh H, Papakonstantinou I, Popoola W O, Burton A and Cacialli F 2015 Wavelength-multiplexed polymer LEDs: towards 55 Mb/s organic visible light communications *IEEE J. Sel. Areas Commun.* **33** 1819–28
[38] Ryan W and Lin S 2009 *Channel Codes: Classical and Modern* (Cambridge: Cambridge University Press)
[39] Shannon C E 2001 A mathematical theory of communication *ACM SIGMOBILE Mob. Comput. Commun. Rev.* **5** 3–55
[40] Cho J, Park J H, Kim J K and Schubert E F 2017 White light-emitting diodes: history, progress, and future *Laser Photonics Rev.* **11** 1600147
[41] Dupuis R D and Krames M R 2008 History, development, and applications of high-brightness visible light-emitting diodes *J. Light. Technol.* **26** 1154–71
[42] Nishiura S, Tanabe S, Fujioka K and Fujimoto Y 2011 Properties of transparent Ce:YAG ceramic phosphors for white LED *Opt. Mater.* **33** 688–91
[43] Elliot A J and Maier M A 2014 Color psychology: effects of perceiving color on psychological functioning in humans *Annu. Rev. Psychol.* **65** 95–120
[44] Popoola W O 2016 Impact of VLC on light emission quality of white LEDs *J. Lightwave Technol.* **34** 2526–32
[45] Le Minh H, O'Brien D, Faulkner G, Zeng L, Lee K, Jung D, Oh Y and Won E T 2009 100-Mb/s NRZ visible light communications using a postequalized white LED *IEEE Photonics Technol. Lett.* **21** 1063–5
[46] Le Minh H, O'Brien D, Faulkner G, Zeng L, Lee K, Jung D and Oh Y 2008 80 Mbit/s visible light communications using pre-equalized white LED *34th European Conf. Optical Communication (EPOC 2008)* (IEEE) pp 1–2
[47] Wang Y, Wang Y, Chi N, Yu J and Shang H 2013 Demonstration of 575-Mb/s downlink and 225-Mb/s uplink bi-directional SCM-WDM visible light communication using RGB LED and phosphor-based LED *Opt. Express* **21** 1203–8
[48] Wu F-M, Lin C-T, Wei C-C, Chen C-W, Huang H-T and Ho C-H 2012 1.1-Gb/s white-LED-based visible light communication employing carrier-less amplitude and phase modulation *IEEE Photonics Technol. Lett.* **24** 1730–2
[49] Wang S-W, Chen F, Liang L, He S, Wang Y, Chen X and Lu W 2015 A high-performance blue filter for a white-LED-based visible light communication system *IEEE Wireless Commun.* **22** 61–7
[50] Li H, Chen X, Huang B, Tang D and Chen H 2014 High bandwidth visible light communications based on a post-equalization circuit *IEEE Photonics Technol. Lett.* **26** 119–22
[51] Sung J-Y, Chow C-W and Yeh C-H 2014 Is blue optical filter necessary in high speed phosphor-based white light LED visible light communications? *Opt. Express* **22** 20646–51
[52] Haigh P A, Ghassemlooy Z, Rajbhandari S, Papakonstantinou I and Popoola W 2014 Visible light communications: 170 Mb/s using an artificial neural network equalizer in a low bandwidth white light configuration *J. Lightwave Technol.* **32** 1807–13
[53] Vučić J, Kottke C, Nerreter S, Habel K, Büttner A, Langer K-D and Walewski J W 2010 230 Mbit/s via a wireless visible-light link based on OOK modulation of phosphorescent white LEDs *Optical Fiber Communication Conf.* (Optical Society of America) p OThH3
[54] Yeh C-H, Liu Y-L and Chow C-W 2013 Real-time white-light phosphor-LED visible light communication (VLC) with compact size *Opt. Express* **21** 26192–7

[55] Sajjad M T *et al* 2015 Novel fast color-converter for visible light communication using a blend of conjugated polymers *ACS Photonics* **2** 194–9
[56] Chun H *et al* 2014 Visible light communication using a blue GaN μLED and fluorescent polymer color converter *IEEE Photonics Technol. Lett.* **26** 2035–8
[57] Manousiadis P *et al* 2015 Demonstration of 2.3 Gb/s RGB white-light VLC using polymer based colour-converters and GaN micro-LEDs *Summer Topicals Meeting Series (SUM), 2015* (IEEE) pp 222–3
[58] Haigh P A, Burton A, Werfli K, Le Minh H, Bentley E, Chvojka P, Popoola W O, Papakonstantinou I and Zvanovec S 2015 A multi-CAP visible-light communications system with 4.85-b/s/Hz spectral efficiency *IEEE J. Sel. Areas Commun.* **33** 1771–9
[59] Wang Y, Shao Y, Shang H, Lu X, Wang Y, Yu J and Chi N 2013 875-Mb/s asynchronous bi-directional 64QAM-OFDM SCM-WDM transmission over RGB-LED-based visible light communication system *Optical Fiber Communication Conf.* (Optical Society of America) p OTh1G–3
[60] Wu F-M, Lin C-T, Wei C-C, Chen C-W, Chen Z-Y and Huang K 2013 3.22-Gb/s WDM visible light communication of a single RGB LED employing carrier-less amplitude and phase modulation *Optical Fiber Communication Conf.* (Optical Society of America) p OTh1G–4
[61] Chen S-H and Chow C-W 2014 Color-shift keying and code-division multiple-access transmission for RGB-LED visible light communications using mobile phone camera *IEEE Photonics J.* **6** 1–6
[62] Wang Y, Tao L, Huang X, Shi J and Chi N 2015 8-Gb/s RGBY LED-based WDM VLC system employing high-order CAP modulation and hybrid post equalizer *IEEE Photonics J.* **7** 1–7
[63] Cui L, Tang Y, Jia H, Luo J and Gnade B 2016 Analysis of the multichannel WDM–VLC communication system *J. Lightwave Technol.* **34** 5627–34

IOP Publishing

Visible Light

Data communications and applications

Paul Anthony Haigh

Chapter 2

Technological enablers

2.1 Introduction

There are a significant number of excellent texts that cover the fine details of inorganic [1–3] and organic [4–7] semiconductors, and hence, only the basic requirements for VLC technologies are covered here. A semiconductor is so-called because it is a medium between a conductor (such as a metal) and an insulator (i.e. rubber). After certain conditions have been satisfied, i.e. an energy barrier is overcome, a semiconductor can pass current and/or generate/absorb light. In general, semiconductors have a positive (*p*)-type and a negative (*n*)-type contact that allows a current to flow through the materials that make up the device. The material that a semiconductor is made of is of crucial importance to its operating characteristics. For example, for blue-light emission from an LED, GaN is normally used as the emissive layer material due to its band-gap energy (E_g), which is 3.4 eV. The band gap simply defines the energy range in which electron states can exist, and is given as the difference between the energy levels at the bottom of the conduction band energy level (E_g) and the top of the valence band energy level (E_v). The valence and conduction bands, in turn, are defined as the following: (i) valence band: the highest band in a material occupied by an electron; and (ii) conduction band: the lowest unfilled energy band in a material. In order for a semiconductor to conduct, the band-gap energy must be exceeded in order to promote an electron from the valence band into the conduction band and form a current flow. This concept is illustrated conceptually in figure 2.1, where E_v and E_c represent the energy level of the valence and conduction bands, respectively. The generalised band-gap energy E_g can be defined mathematically as follows:

$$E_g = E_c - E_v \tag{2.1}$$

Given a semiconductor with a 3.4 eV band gap such as GaN, the expected emission wavelength can easily be calculated using the following relationship [1]:

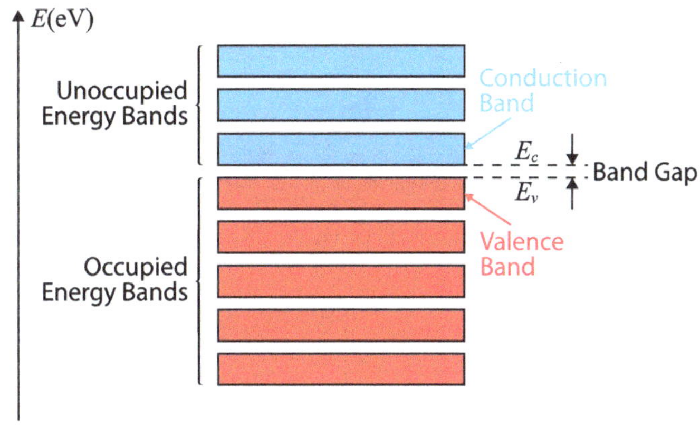

Figure 2.1. Illustration of the conduction and valence band of a simple semiconductor.

$$\lambda(\text{nm}) = \frac{hc}{E_g} = \frac{1.24 \times 10^{-6}}{E_g} \tag{2.2}$$

where h is the Planck constant, given as $h = 4.136 \times 10^{-15}$ eV s, and λ is the wavelength. Therefore, when $E_g = 3.4$ eV in the case of GaN, $\lambda = 364.7$ nm, which is approximately blue in colour. For red or green wavelengths, one may select GaP (1.6–2.03 eV) or InGaN (1.9–4.0 eV) as the material. Other important and relevant materials are Si and indium gallium arsenide (InGaAs), which are both used extensively for PDs. As a general rule-of-thumb, the conductivity of a device increases with a decreasing band gap, as less energy is required to overcome the forbidden zone between the valence and conduction bands. Conductors are often found with little or no band gap (<0.1 eV), while insulators are generally very large (>3 eV) [1].

2.2 Principles of light generation and absorption

Light is made of photons, which are elementary particles with zero mass that move at the speed-of-light c, which is 299 792 458 m s^{-1} in a vacuum. In order to provide illumination, LDs and LEDs must generate photons and they do that in two distinct ways. The former generate photons using a method called stimulated emission that can be referred to in the literature [1–3] as it is not covered here. The latter uses a process called spontaneous emission, which is based on directly converting electron/hole pairs into photons through recombination at the emissive layer of the LED.

The simplest structure that can enable light generation is the p–n junction that consists of a p-type contact that injects holes into the device, an n-type contact that injects electrons, and a depletion region between them. When this structure is forward-biased with a voltage, electron–hole recombination occurs when the band-gap energy is overcome and an electron is promoted into the conduction band and recombines with a hole, resulting in a single photon emission. If the device is reverse-biased, when a photon of energy in excess of the band gap impinges on the surface, a charge carrier is generated and the electron and hole that are produced will be

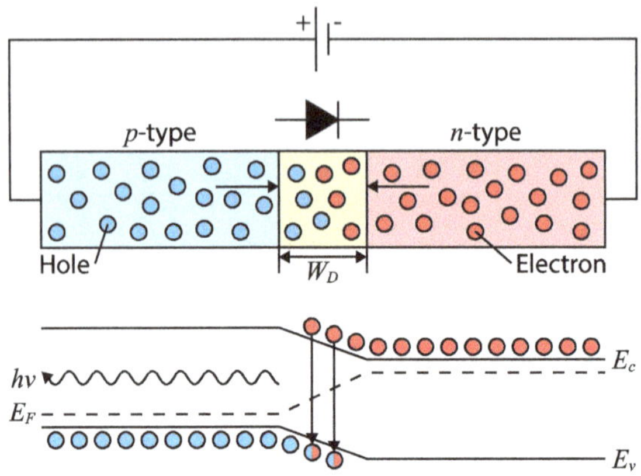

Figure 2.2. A simple *p–n* junction that forms the basis of LED technology and is the main technological enabler that supports VLC technologies.

attracted to their respective *n*- and *p*-type electrodes, inducing a photocurrent. These two processes are called radiative recombination and will be expanded upon in the next subsection. A *p–n* junction is illustrated in figure 2.2 for light emission, and it should be noted that the Fermi level (E_f) is not covered here but further elaboration can be found in [1]. The region over which the holes and electrons diffuse is called the depletion layer and has width W_D.

2.2.1 Radiative recombination of electrons and holes

When a *p–n* junction is absent of a bias voltage, the concentration of electrons n_0 and holes p_0 are equivalent to that of the intrinsic carrier concentration n_i at a given temperature, following [8]:

$$n_i^2 = n_0 p_0 \tag{2.3}$$

The recombination rate R of charge carriers follows [3, 8]:

$$R = Bnp \tag{2.4}$$

where n and p are the overall carrier concentrations and B (cm^3 s^{-1}) is the bimolecular recombination coefficient. This condition is only valid when the junction is free from external electrical or optical carrier injection. However, when a bias is applied, the overall concentration of charge carriers deviates as [3, 8]:

$$n = n_0 + \delta n \tag{2.5}$$

$$p = p_0 + \delta p \tag{2.6}$$

where δn and δp are the excess charge carriers induced by the bias voltage and it is not necessary that $p = n$ for correct operation.

When an external bias is applied, (2.4) is modified as follows [3]:

$$R = B(np - n_0 p_0) \tag{2.7}$$

Now, the concept of low-level injection is introduced, which simply implies [3]:

$$(n_0 + p_0) \gg \delta n \tag{2.8}$$

$$(n_0 + p_0) \gg \delta p \tag{2.9}$$

for either type. Further assuming that $\delta n = \delta p$ because generation/recombination occur in pairs, the recombination rate can be redefined as [3, 8]:

$$R = B(np - n_0 p_0) \tag{2.10}$$

Through substitution of (2.8) and (2.9) and expansion, the following emerges [3]:

$$R = B(p_0 + n_0 + \delta n)\delta n = \frac{\delta n}{\tau} \tag{2.11}$$

where τ is the radiative lifetime and recalling that $\delta n = \delta p$. Therefore the radiative lifetime is given as [3, 8]:

$$\tau = \frac{1}{B(n_0 + p_0 + \delta n)} \tag{2.12}$$

Hence, R is directly proportional to the charge carrier concentrations, which in turn are impacted by the bias voltage. The bimolecular recombination coefficient is given by [8]:

$$B = 3 \times 10^{-10} \left[\frac{300}{T}\right]^{3/2} \left[\frac{E_g}{1.5}\right]^2 \tag{2.13}$$

where T is the temperature.

Now, the concept of high-level injection is introduced, which is the opposite condition to (2.8) and (2.9), i.e.:

$$\delta n \gg (n_0 + p_0) \tag{2.14}$$

$$\delta p \gg (n_0 + p_0) \tag{2.15}$$

and hence, the recombination rate for high-level injection, following the same procedure as previously, becomes:

$$R \simeq B\delta n^2 \simeq Bn^2 \tag{2.16}$$

This condition is known as the spontaneous recombination rate and is the associated process with modern high powered LEDs and LDs, the latter of which are also subject to stimulated emission of photons, which is where a photon with energy above the band gap induces an electron–hole recombination, resulting in an additional photon emission. Stimulated emission is not covered here but may be referred to in the literature [1, 3].

Generally, the carrier lifetime is not controlled in VLC systems since the LEDs used are commercial devices and not bespoke for communications applications.

Therefore, there has been a significant effort to provide models of *p–n* junctions that accurately describe their behaviours, as will be seen in the next subsection.

2.2.2 Light-emitting diodes

There are a set of well established relationships that clearly define semiconductor diode behaviour in terms of their current and voltage. The first of which is known as the Shockley diode equation and was proposed by William Shockley in 1949 during his time at Bell Labs [9]. Shockley was later part of a team of three scientists, including John Bardeen and Walter Brattain who were awarded the 1956 Nobel Prize for Physics due to their pioneering contributions to the field of semiconductors and transistors. The Shockley equation is given as follows [9]:

$$I = I_0 \exp\left[\frac{qV_B}{k_B T} - 1\right] \tag{2.17}$$

where I is the current in the diode, I_0 is the saturation current density (refer to [1] for the mathematical relationship and complete derivation), q is the charge of an electron, which is commonly accepted as 1.602×10^{-19} C, V_B is the bias voltage, T is the temperature (K), and k_B is the Boltzmann constant, given by 1.38×10^{-23} m^2 kg s^{-2} K^{-1}, or J K^{-1}.

In figure 2.3, the Shockley diode equation is illustrated for both the forward and reverse bias operating regions. Note that in reverse bias, the diode will have some leakage current I_0, which should be as low as possible in ideal circumstances. Note also, that the relationship passes through the origin and the reason for this is due to Ohm's law; if voltage is absent then so is current.

As was mentioned in chapter 1, one of the main considerations in VLC systems is the low bandwidths exhibited by the LEDs used. The root cause of the bandwidth is related to the depletion region of the *p–n* junction, which has a contribution to the

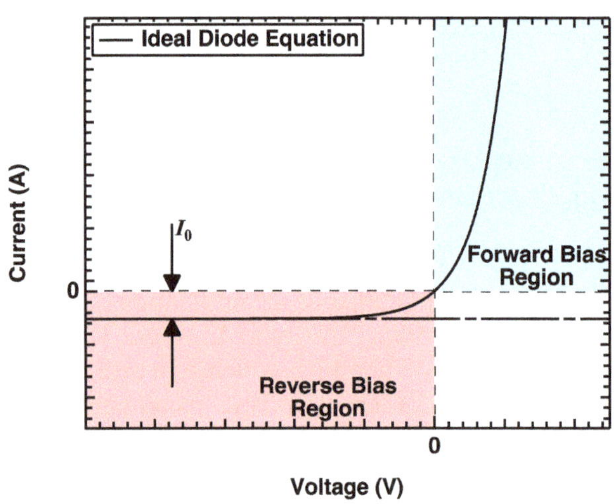

Figure 2.3. The ideal diode equation, following the Shockley equation (2.17) and indicating the forward and reverse bias regions.

equivalent plate capacitance of the diode. The width of the depletion region is given as the following [8]:

$$W_D = \sqrt{\frac{2\varepsilon_0\varepsilon_r\left(N_A^+ + N_D^-\right)}{qN_A^+N_D^-}(V_{bi} - V_B)} \tag{2.18}$$

The dielectric constant of the depletion region material is given by ε_r. the relative permittivity of a vacuum is given by ε_0 (8.854×10^{-12} $\mathrm{Fm^{-1}}$), N_A^+ and N_D^- are the concentrations of the electron acceptor and donor, respectively and finally, V_{bi} is the built-in diffusion voltage which is defined by [10]:

$$V_{bi} = \frac{k_BT}{q}\ln\left[\frac{N_A^+N_D^-}{n_i^2}\right] \tag{2.19}$$

recalling that n_i is the intrinsic carrier concentration at thermal equilibrium.

Allow the assumption that the acceptor/donor concentrations are constant at any given time instant, and also recalling that ε_r, ε_0 and q are constant and pre-defined, it is clear that the width of the depletion region is controlled by the difference between the built-in diffusion voltage (also fixed with the previous presumptions) and the bias voltage. If the bias voltage is positive (forward bias), then the width of the depletion region reduces and it is easier for charge carriers to transverse the forbidden energy region as less energy is required to promote an electron into the conduction band, which is clearly a desirable condition for LEDs. The opposite condition occurs when the bias voltage is negative (i.e. reverse bias), and the width of the depletion region increases, meaning that more energy is required for charge carriers to diffuse, which is clearly a good option for preventing leakage current in PDs.

The width of the depletion region has an associated capacitance which relates to the width through the following [10]:

$$C_j = \sqrt{\frac{\varepsilon_0\varepsilon_r qN_A^+N_D^-}{2\left(N_A^+ + N_D^-\right)\left(V_{bi} - V_B\right)}} = \frac{\varepsilon_0\varepsilon_r}{w} \tag{2.20}$$

When considering dimensionality, (2.20) can be re-written as:

$$C_j = \frac{\varepsilon_0\varepsilon_r A}{d} \tag{2.21}$$

where A is the cross-sectional area and d is its thickness.

The relationship between junction capacitance and depletion region width is illustrated in figure 2.4, where several example thicknesses are used as an example. The dielectric constant is taken from [11] as 5.35 at 300 K.

Equation (2.21) is a key relationship in VLC, because it is the fundamental limit to the switching speed of the diode and therefore defines the overall bandwidth. Since the charge carriers have a finite lifetime, it is clear that diodes such as the *p–n* junction outlined above and LEDs have a low-pass transfer function in the

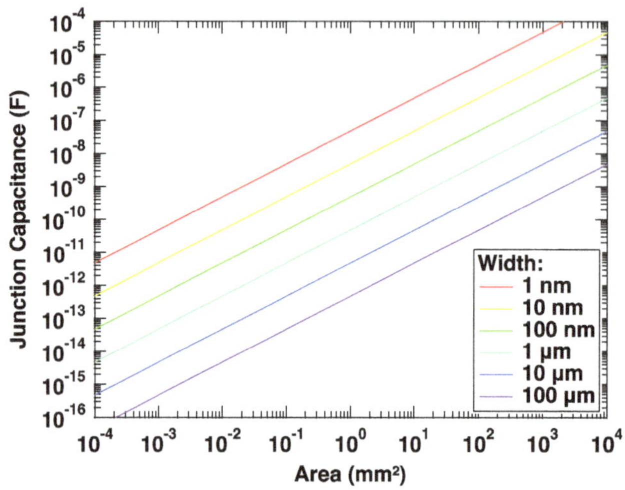

Figure 2.4. The junction capacitance as a function of cross-sectional area and width. Clearly, as the device gets physically large, the junction capacitance increases resulting directly in a reduction of useful bandwidth.

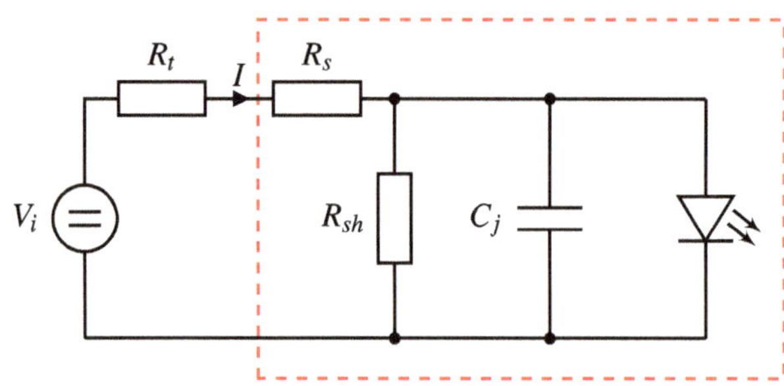

Figure 2.5. An equivalent circuit for an LED including contact resistance (R_s), bulk resistance (R_{sh}) and junction capacitance.

frequency domain and hence the following is also true for the majority of LEDs used in VLC:

$$f_c = \frac{1}{2\pi R_s C_j} \tag{2.22}$$

where f_c is the low-pass cut-off frequency and R_s is the series resistance applied to the diode. Thus, through combination of (2.21) and (2.22), the overall relationship becomes:

$$f_c = \frac{d}{2\pi\varepsilon_0\varepsilon_r R_s A} \tag{2.23}$$

An equivalent circuit of an LED is shown in figure 2.5, where R_{sh}, R_s and R_t are the shunt, series and input resistances of the diode, respectively, and V_i is the input voltage. The diode on the right-hand side can be considered as ideal.

As previously discussed, ε_0, ε_r are fixed and the dimensions of the depletion region depend on the external bias voltage, which in turn control the equivalent capacitance. Hence, by adjusting the series resistance of the diode, which is selected by the circuit designer, the bandwidth can be controlled within a certain range, depending on its physical characteristics, as is illustrated in figure 2.6 using (2.23) with the following fixed parameters: $\varepsilon_r = 5.35$ and $d = 1$ nm with the same range of A as previously, and a varying R_s. Figure 2.6 shows that reducing the series capacitance to a few Ohms can significantly increase the device's low-pass cut-off frequency, which is a substantial advantage when considering high speed communications systems. This will form the basis of the discussion of VLC circuit design in chapter 3.

2.2.3 Photodetectors

The operating principles of PDs are the same as discussed in the previous sections, however, in reverse bias. The objective of a PD is to generate a photocurrent that accurately describes the incident light that impinges on its photoactive area. PDs are square-law detectors, which means that the electrical current generated is proportional to the square of the optical field. Hence, the generated photocurrent I_p is proportional to the strength of the optical power $_{Pi}$ at the receiver, following [12]:

$$I_p = \eta \frac{qP_i}{h\nu}[1 - e^{-\alpha l}] \tag{2.24}$$

where η is the quantum efficiency of the device, i.e. the ratio of generated charge carriers that contribute to the photocurrent, α is the absorption coefficient of the

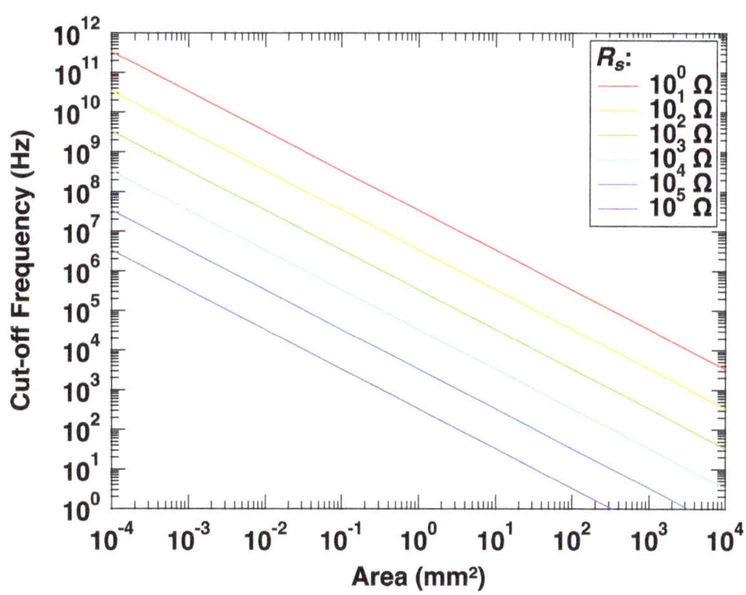

Figure 2.6. Cut-off frequency as a function of contact resistance and cross-sectional area; as expected the device bandwidth reduces with increasing resistance and area.

photoactive material and measures the fraction of optical power absorbed across a unit length, l of the aforementioned material. The photon frequency is given by ν m s^{-1}. Typically for VLC systems, the detector material is Si, which is inexpensive and one of the few materials with relatively good performance in the visible range. For IR and NIR applications, typically InGaAs is preferred due to its high responsivity.

The responsivity is an important figure-of-merit for PDs because it defines the ratio of incident photons to generated photocurrent and hence, high responsivity is desirable for PDs in VLC systems because it inherently leads to higher SNRs. Responsivity $\Re$ (A W^{-1}) is defined by the following [12]:

$$\Re = \frac{I_p}{P_i} = \frac{\eta q}{h\nu}[1 - e^{-\alpha l}] \tag{2.25}$$

Ideally, η and the term $1 - e^{-\alpha l}$ would approach unity and hence, responsivity would be defined by $qh^{-1}\nu^{-1}$, although there are physical limitations that restrict this. The responsivities of several popular semiconductors are shown in figure 2.7, where some new materials are introduced including germanium (Ge), which has not been used in VLC to date because it is not a visible light absorber, and poly(3-hexylthiophene-2,5-diyl) (P3HT), an organic photodetector (OPD) that was first introduced for VLC systems in [13].

The polymer P3HT will be discussed further in a later subsection; however, it offers several clear advantages in comparison to Si from observation of figure 2.7. The first of which is the higher responsivity, which is around four times higher at blue wavelengths (400 nm) meaning that a higher proportion of the light collected is translated into photocurrent, and secondly, the cut-off wavelength is shorter, meaning that no IR light is absorbed, meaning that out-of-band noise is avoided to a certain extent. Finally, the cut-off wavelength can be modified, that is made longer to absorb NIR wavelengths if

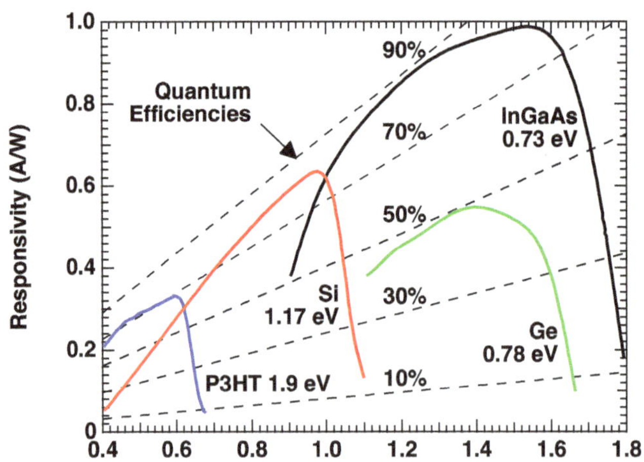

Figure 2.7. Responsivities of a number of different PD materials. The main ones that are used in VLC are silicon (Si) and poly(3-hexylthiophene-2,5-diyl) (P3HT).

desirable by doping of the polymer with other materials [14]. The cut-off wavelength of a PD is defined by the band-gap energy E_g, as mentioned, because the incoming photon must have sufficient energy to overcome it, as was defined in (2.2).

In contrast to LEDs, the PDs used in VLC cannot be considered as *p–n* junctions directly, because their RC_j time constants are excessive and absorption is low in the diffusion lengths [3]. Instead, a wide intrinsic layer is inserted between the *p*-type and *n*-type materials as the major absorber. These structures are called positive-intrinsic-negative (PIN), and their modified structure is shown in figure 2.8, where R_L is a load resistor.

The Shockley equation from (2.17) remains valid for PDs. Two *V–I* curves can be seen for an example PD in figure 2.9, where two curves can be seen. One is for the dark leakage current, which is the residual current that flows through the PD constantly, even when there is no illumination, and the second is the generated photocurrent. The difference between them is clearly I_p. Recalling that a reverse bias voltage V_B is required to deplete the active region of charge carriers, it is desirable to increase V_B as far as possible to increase the intrinsic region width and hence, reduce the equivalent capacitance, resulting in a higher device bandwidth.

The bandwidth is also mainly limited by the capacitance as has been previously demonstrated. The reason for this is because the PIN structure can largely be considered approximately equivalent to the *p–n* junction albeit with a larger depletion region (the intrinsic layer) and hence, the mathematics of the charge carrier transport stays constant [3]. The equivalent circuit is shown in figure 2.10.

The PD behaves as a current source when illuminated. When operated in reverse bias, junction capacitance is minimised as far as possible, depending on the intrinsic region width and bias voltage as previously described. The photocurrent is generated from three separate sources, the impinging illumination I_p, the dark current I_d and the noise current I_n. The sum total of these currents is then distributed across the resistors. Typically the shunt resistance R_{sh} is in the order of several MΩ and the series resistance R_s is relatively low in comparison. The load resistance R_L defines V_o and is selected by the circuit designer. The diode shown on the left-hand side can be considered ideal. The output voltage can be obtained by inspection of the circuit as follows:

$$V_o = [I_p + I_d + I_n][R_L R_{sh}][R_L + R_{sh} + R_s] \tag{2.26}$$

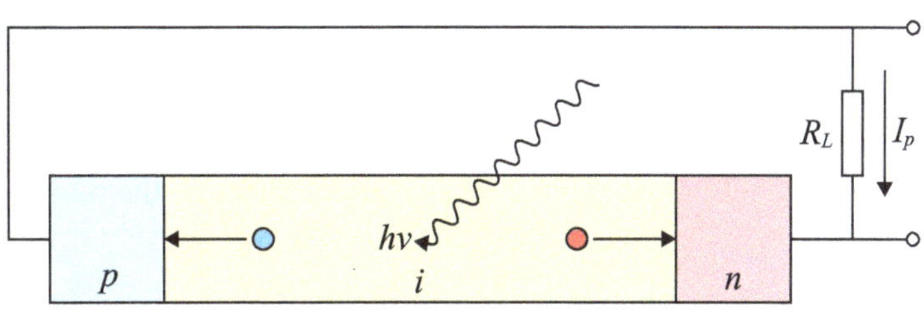

Figure 2.8. A simple PIN structure complete with intrinsic region for absorption of photons before separation into electrons and holes.

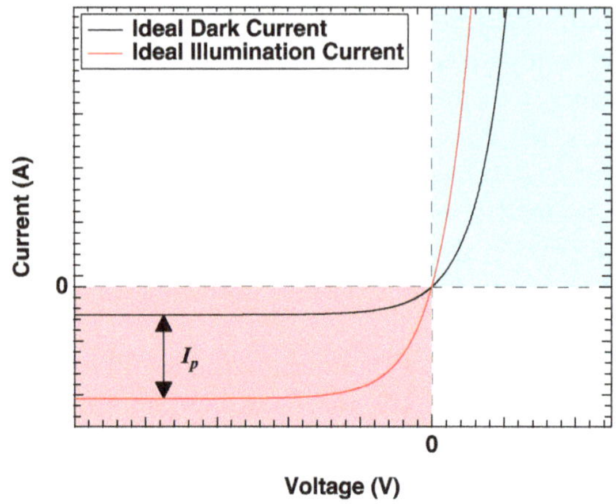

Figure 2.9. The Shockley equation once more, however, dark current is now introduced, which is the residual current that flows through the PD even when no light is incident. The dark current is a source of noise.

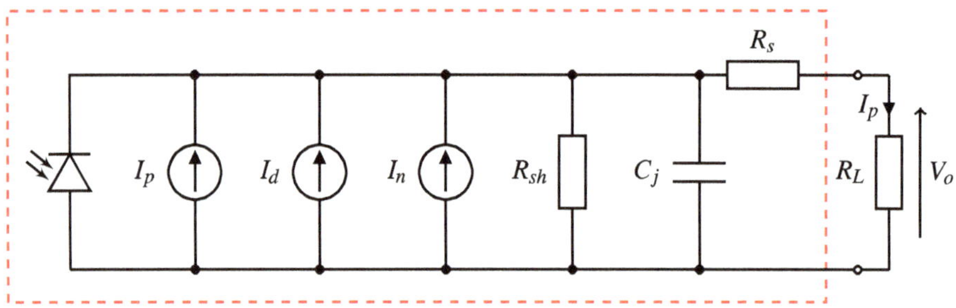

Figure 2.10. The equivalent circuit model for a PD, contained within the red dashed box.

2.2.4 Noise current

The noise equivalent power (NEP) is an important concept for PDs because it defines the minimum detectable optical power for a given device. The NEP is defined as the ratio of the minimum incident optical power on to the active area of the PD required to generate a photocurrent equal to that of the total noise current, i.e. the power required to give a unity SNR in 1 Hz of bandwidth, as follows [15]:

$$NEP = \frac{S_n}{\Re} \tag{2.27}$$

where S_n is the noise spectral density and recalling that $\Re$ is the responsivity, resulting in a unit of $\mathrm{W}/\sqrt{\mathrm{H}}\mathrm{z}$. Clearly the NEP depends on the wavelength, since the responsivity is wavelength dependent. For any given detector, the minimum NEP is given when the responsivity is at its maxima. This is useful because it can be used to calculate the noise at any operating wavelength of the PD using the following relationship [16]:

$$NEP(\lambda) = NEP_{\min}\left[\frac{\Re_{\max}}{\Re(\lambda)}\right] \tag{2.28}$$

and hence, the minimum detectable power at a given wavelength is defined as [16]:

$$P_{\min} = NEP(\lambda)\sqrt{B} \tag{2.29}$$

where B is the measurement bandwidth. The specific detectivity (D^*) of a PD is the reciprocal of the NEP, normalised to the square root of the photoactive area of the diode:

$$D^* = \frac{\sqrt{A}}{NEP} \tag{2.30}$$

There are several sources that combine to produce the overall noise current I_n. Two of the most important are shot noise and Johnson (thermal) noise. Shot noise is primarily the leakage current of a PD under reverse bias and its variance σ_{shot}^2 is given by the following [17]:

$$\sigma_{\text{shot}}^2 = 2\Re q P_i B + 2qI_{\text{bg}}B \tag{2.31}$$

Clearly, from (2.31), where the left-hand side is the leakage current, there is another term, on the right-hand side. This is the introduction of additional shot noise caused by generation of an additional current, the background current I_{bg}, which is introduced via ambient photons incident to the diode that do not form part of the signal and occur from other sources. On the other hand, Johnson noise is also important and is generated in the receiver electronics, and is often the most dominant source in a VLC system. The Johnson noise variance $\sigma_{\text{thermal}}^2$ is given by [17]:

$$\sigma_{\text{thermal}}^2 = \frac{4kTB}{R_L} \tag{2.32}$$

and the total noise variance σ_{t}^2 at the receiver is given by the sum of the shot and Johnson noise [17]:

$$\sigma_{\text{t}}^2 = \sigma_{\text{shot}}^2 + \sigma_{\text{thermal}}^2 \tag{2.33}$$

This concludes the introduction to inorganic semiconductors and their applications in VLC. The next section will introduce organic semiconductor technologies in analogy to their inorganic counterparts. The motivation for this is due to the ever increasing popularity of organic semiconductors in VLC systems.

2.3 Organic semiconductors

The difference between organic semiconductors and inorganic ones is simply the materials used. In organics, the core materials are primarily carbon and nitrogen [18]. They are considered to be a sub-domain of semiconductor in isolation because they possess several of the individual characteristics of metals, such as conductivity with sufficient input energy, and plastics, as they can offer mechanical flexibility. Other advantages over inorganics include wet processing methods, avoiding the use of expensive vacuum or high temperature methods that produce crystalline structures, the ability to produce large, arbitrarily shaped photoactive areas and the enormous potential to provide extremely large scales of production at ultra low costs. All of these characteristics make organics an extremely attractive area for future electronics and optoelectronics industries, which are expected to be valued at around $330B by 2027, which is higher than the Si market value today ($225B) [19]

A number of new technologies have been developed that take advantage of organics including OLEDs [20], OPDs [21], organic photovoltaics (OPVs) [22] and flexible electronics [23], and even IoT applications using organic RF antenna [24]. For instance, in [23], an organic microprocessor was reported based on transistors that offers a full 8-bit programmable interface based on 3381 transistors. This is incomparable to current inorganic microprocessors, which have billions of transistors-per-chip. On the other hand, the first Si microprocessors only offered 4004 transistors on chip, which is comparable [25]; however, the power consumption is reduced comparatively by five orders of magnitude [23]. In figure 2.11 conductivity ranges for common polymers used in organic applications are shown in comparison to gallium arsenide (GaAs) and Si. Polymers are created by reactions between neighbouring atoms, which cause energy to be released and mixing between molecular orbitals, which is a process known as hybridisation [18, 26].

2.3.1 Hybridisation

The process of hybridisation is the underlying mechanism that causes the bonds between molecules. After polymer formation, defining the individual position of any arbitrary single atom is impossible and hence, it is more appropriate to define delocalised electron densities. The bonds between these densities electrons are called π-bonds and they are weak due to their electron dislocation. Other atoms that can be accurately located form σ-bonds which are significantly stronger than π-bonds. Clearly, it is easier to free an electron from a π-bond than a σ-bond and this forms the basis of semiconducting in organics [18, 26].

2.3.2 Linear combination of molecular orbitals

The bonds mentioned so far only take into account individual atoms. Molecules are groups of atoms held together by chemical bonds, and it is necessary to understand the behaviour of molecules as they approach one another. Before one can gain this understanding, it is first important to discuss linear combination of atomic orbitals (LCAO), which is where two atomic orbitals form a an overall molecule. Rudimentary knowledge of orbitals is presumed, but the literature [18] demonstrates that there are several orbital shapes of interest for the majority of organic semiconductors, including the $1s$, $2s$, $2p_x$, $2p_y$ and $2p_z$ orbitals. Other orbitals include the d and f types. The different letters indicate different orbital shapes, as can be seen in figure 2.12.

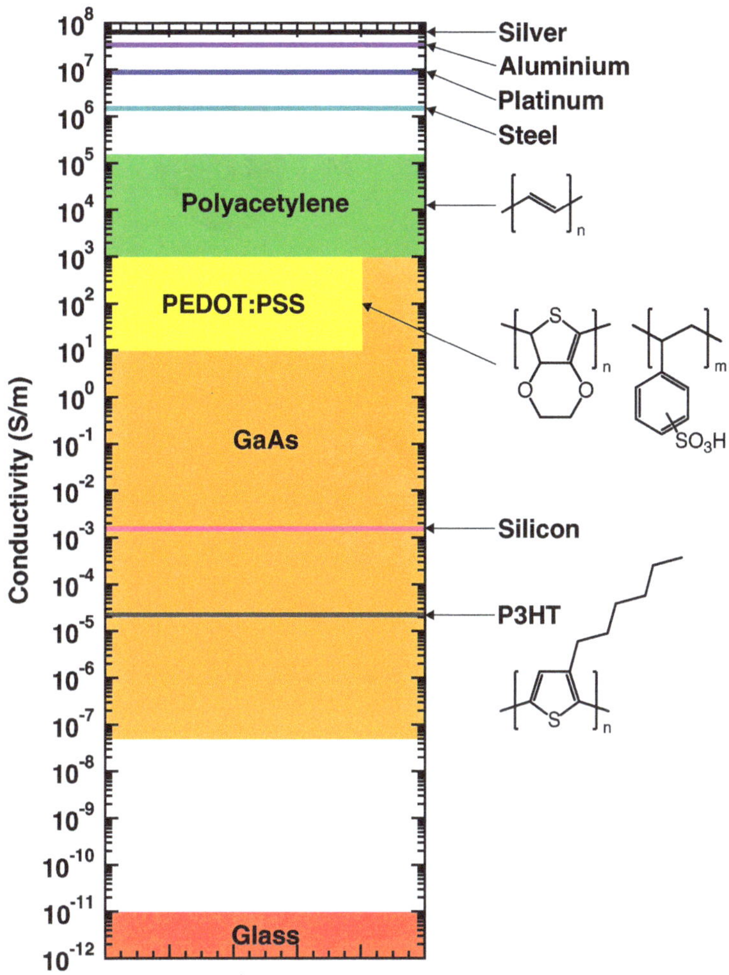

Figure 2.11. Conductivity of some common organic polymers in comparison to common metal conductors and electrical insulators.

The commonly accepted Bohr model of the atom shows that electrons circle the nucleus; however, this is not the case, as electrons can be probabilistically found at any location in a given orbital, depending on the particular element. The numeric precursor simply describes the order in which the orbitals appear in terms of how close they are to the nucleus in terms of energy level. The superscript index describes how many electrons fill that orbital. For example, consider a hydrogen atom; in its first orbital one atom can be present and hence the orbital is defined as $1s^1$. Hydrogen only has one electron and hence cannot fill the $1s$ shell, however helium has two, and as such the same shell is fully occupied and the orbital is denoted as $1s^2$. It should be noted that an s orbital always has slightly less energy than a p orbital for a given energy level, and hence the s orbital always fills before the corresponding p.

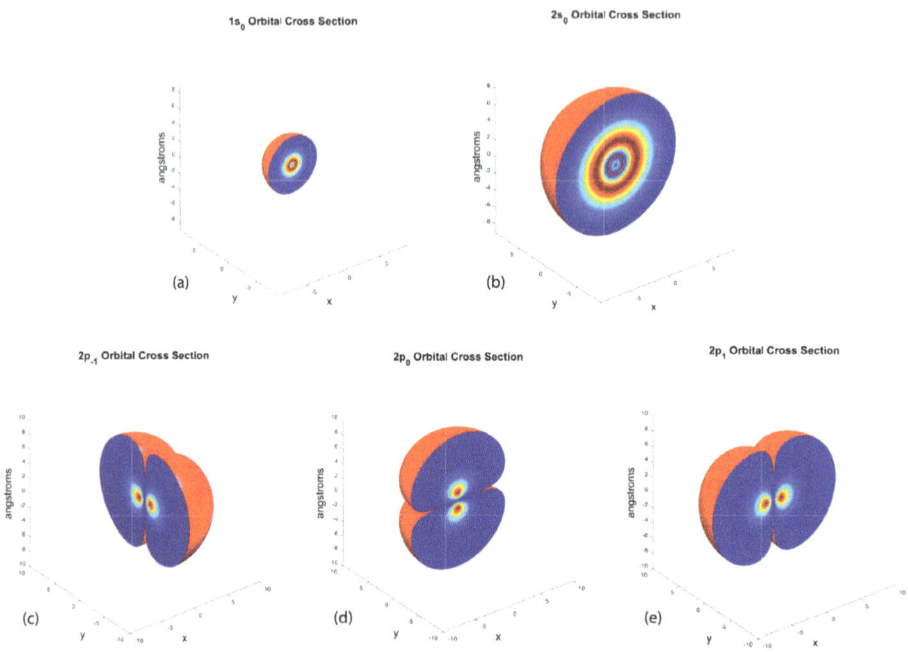

Figure 2.12. Orbital shapes for the first two *s* orbitals and the first *p* orbital.

The *s* orbitals are spherical in shape, while the *p* orbitals have directionality and orthogonality to each other, as can be seen in figure 2.12.

When an atomic orbital forms it has a phase in the same manner as a standing wave. Combination of out-of-phase orbitals produce an anti-bonding orbital while combination of in-phase orbitals produces a bonding orbital. As may be expected, anti-bonding orbitals contain no electrons between the nuclei of the adjoined atoms. This aids the repulsion mechanism between positively charged nuclei, and hence, raises the relative energy level in comparison to the unbounded molecules previously discussed. On the other hand, an in-phase bonding orbital has electrons between the two atomic nuclei, and hence, a drop in their relative energy level can be found because the electrons are shared. Hence, discrete energy levels can be found as illustrated in figure 2.13, and this is a σ-bond as the discrete electron locations can be known.

In the event that two atoms with electrons in the *p* shell meet, both σ- and π-bonds are generated, as is shown in figure 2.14, where * indicates the anti-bonding combination. It is clear from figure 2.14 that less energy is required to promote an electron between π-bonds than between σ-bonds.

If molecules approach and combine that are not of the same type, a bond will still be formed; however, the two molecules will mix and form a hybrid molecular orbital that retains the same number of orbitals as before the mixing process. For instance, combining an *s* orbital with the *p* orbitals results in a hybrid sp^3 orbital, and its shape is determined by the constructive and destructive superposition of the standard atomic orbitals and their angles [18, 26, 27]. The most common type of hybrid

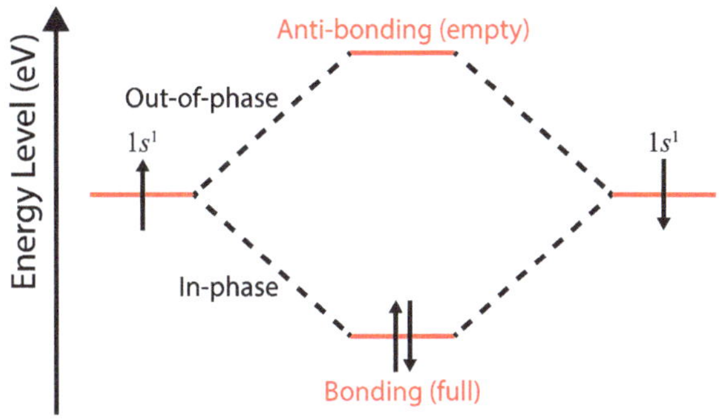

Figure 2.13. Energy level diagram of bonding between two hydrogen atoms, electron spin is indicated by the arrows.

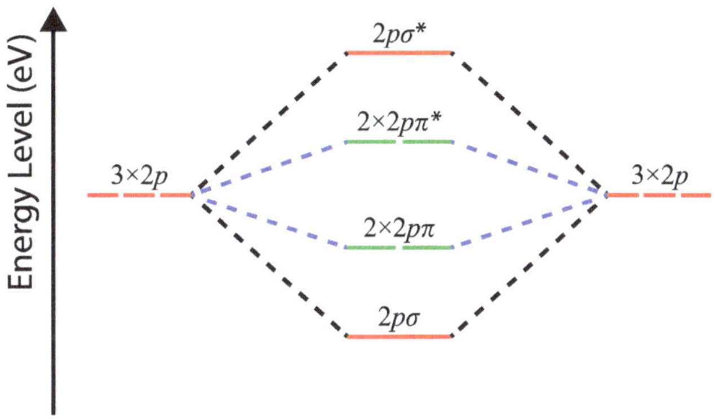

Figure 2.14. Energy level diagram of π and σ bond generation by linear combination of molecular orbitals.

orbital in organic semiconductors is sp^2 and sp^3. Generally, π-bonds are responsible for current conduction through the polymer due to their complete delocalisation across the entire bond chain. As a result, the band-gap energy increases with an increasing number of bonds, until the long chain limit is reached at approximately 20–30 bonds. Therefore, the band-gap energy is modified according to [18]:

$$E_g = E_{LC} + \frac{k_p}{N_\pi} \tag{2.34}$$

where E_{LC} is the long chain limit energy gap, N_π is the number of π-bonds present in the chain, and k_p is a proportionality constant that is introduced to maintain agreement between the theoretically and experimentally observed energy gap relationships [18]. Normally, the energy gap of polymers is approximately 1–4 eV, which nicely and conveniently includes the entire visible range of wavelengths.

2.3.3 LUMO and HOMO

The concept of conduction and valence bands cannot be perfectly introduced into organic semiconductors due to the atomic dislocation of the charge carriers. Similar concepts can be introduced called the lowest unoccupied molecular orbital (LUMO), that is the lowest energy level bond in which no electrons exist. Similarly, the highest occupied molecular orbital (HOMO) is opposite; the highest energy level bond in which electrons are contained.

Generally, the LUMO and HOMO are denoted in the literature by π^* and π, respectively. Clearly, the band-gap energy is the difference in energy between the two, and hence, for photon emission, an electron must be promoted from the HOMO to the LUMO, akin to the valence/conduction bands in inorganics. In the same manner, the emission wavelength corresponds to the energy gap, while for absorption, the impinging photon energy must exceed that of the band gap, as previously described. However, the generation and recombination processes are not equivalent, because charge carriers cannot freely move back to their respective electrodes. As a result, the concept of excitons must now be introduced, which are a quasi-particle that represents a bound electron–hole pair that exists only for (typically) several nanoseconds before photon emission.

In order to separate the exciton, an energy greater than the binding energy must be applied for both generation and recombination. As a result, the literature has shown that there are several variations of the generic exciton that are classified according to their binding energy, (i) Frenkel excitons, (ii) charge transfer excitons and (iii) Mott–Wannier excitons. The most abundant in organics are Frenkel excitons; however, all three are not discussed in detail here, but can be referred to in the literature [18, 28–30]. The Frenkel exciton can move through the polymer chain freely. As one might expect, conductive polymers are not crystalline structures as in metallic semiconductors, and instead consist of a localised density of states (DOS). The positions and energy level of each possible discrete site within the density can be described as a Gaussian distribution [4, 18, 31], given by the following [4, 32]:

$$DOS(E) = \frac{N_t}{\sigma\sqrt{2\pi}} \exp\left[-\left(\frac{E - E_0}{\sigma\sqrt{2\pi}}\right)^2\right] \tag{2.35}$$

The centre of the Gaussian distribution is given by E_0, N_t is the total number of effective DOS and σ is the variance of the DOS. This concept is illustrated in figure 2.15, where the movement of electrons and holes are highlighted by 1a and 1b for the LUMO and HOMO, respectively. The majority carrier in organics are holes, which occupy every single vacant state. Electrons are shown in the LUMO for simplicity, and vice-versa for holes in the HOMO. In section 2, trap states are indicated inside the dashed boxes for holes and electrons. Charge traps have a negative effect as they restrict charge transport properties and restrict the generation and recombination processes as will be discussed in the following paragraphs. Finally, in section 3, a Frenkel exciton is generated and a photon is emitted with energy approximately equivalent to the energy gap of the semiconductor. For absorption of a photon, the opposite process occurs. Starting at section 3, if a photon with energy greater than the band gap impinges on the photoactive area a Frenkel

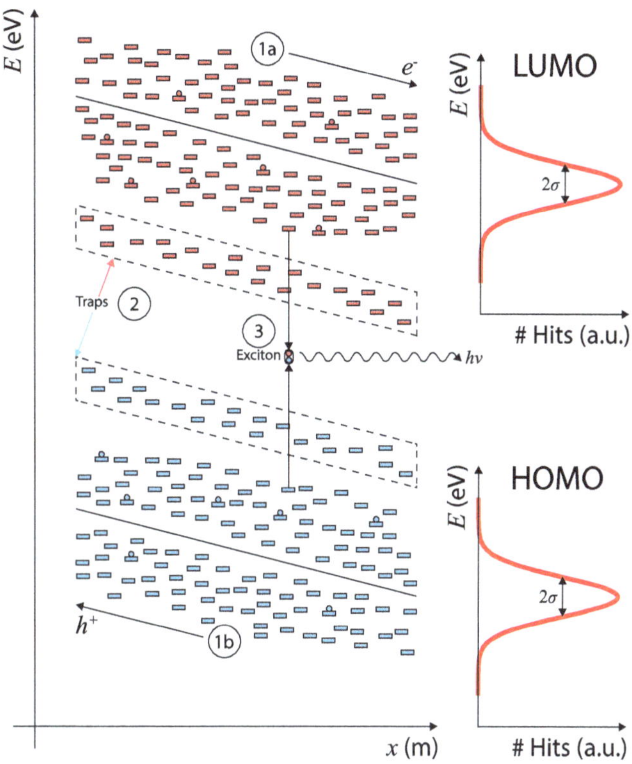

Figure 2.15. Density of states for an organic semiconductor and the processes for photon emission with respect to the HOMO and LUMO levels; electrons and holes hop through localised states in the direction of the e^- and h^+ arrows in sections 1a and 1b, respectively. Trap states exist and the electrons and holes must avoid being restricted in these states to recombine to form a photon. In section 3, a Frenkel exciton is generated and finally, a photon is emitted.

exciton is generated and gets separated into an electron and a hole which are attracted to their relative electrodes through sections 1a and 1b.

Alternatively to photon generation, the carriers can fall into the charge traps highlighted in section 2 to restrict the generated photocurrent. The mobility of charge carriers is perturbed by the addition of trap states at the interfaces (intrinsic traps) and due to defects in the layers and production processes (extrinsic traps). Both types of trap impede transport properties of organics as they interfere with the recombination and generation processes by introducing new energy levels that have energy levels between the HOMO and LUMO levels, meaning charge carriers must drop into the traps as they propagate across the device [33] (figure 2.15). Clearly if a charge carrier is confined inside a trap, it cannot assist with the overall generation of photons or photocurrent. The issue of charge trapping is one of the key challenges facing physical chemists who are researching organics. Charge carriers (and therefore current flow) can be considered to be 'hopping' from state-to-state in the localised densities previously mentioned. The hopping rate υ_{ij} between localised states i and j is known as the Miller–Abrahams rate [34]:

$$v_{ij} = v_0\,(= 2\gamma r_{ij})\begin{cases} \exp\left(\dfrac{E_j - E_i}{k_B T}\right) & \text{if } E_j \geqslant E_i \\ 1 & \text{if } E_j \leqslant E_i \end{cases} \tag{2.36}$$

where v_0 is a pre-factor, γ is the inverse localisation length, r_{ij} is the distance between localised states and $E_j - E_i$ is the energetic barrier that the charge carrier must overcome (as in figure 2.15).

2.3.4 The bulk heterojunction

In terms of OPDs, in order to separate excitons into individual charge carriers, a larger energy is required than inorganics due to the high binding energy as mentioned in the previous subsection.

To offer a solution to this problem, the concepts of electron donor and electron acceptor must be introduced. Electron donors have lower electron affinity (the difference between the band edge and the vacuum energy) than electron acceptors. It should be noted that a high electron affinity is desirable for electron acceptors and vice-versa for electron donors. It is possible to disassociate the exciton at the interface of an electron donor/acceptor configuration due to the unbalanced electron affinities (i.e. the unbalanced energy levels). In this thesis only radiative decay of excitons is considered (non-radiative decay is also possible as previously) and with the influence of an external voltage the generated electrons and holes are attracted to their respective electrodes. The bulk heterojunction (BHJ) is an interpenetrated blend of electron donor and electron acceptor that provides such an interface that is distributed across the entire photoactive area of the organic photonic device as illustrated in figure 2.16. The reason for interpenetration is due to the fact that radiative decay of excitons depends on the distance of exciton generation from the electron acceptor–electron donor border (for distances >10 nm, the exciton will not offer radiative decay [18]). BHJs were first introduced in [21] and are popular in OPVs and OPDs due to the fact that they are soluble and provide extremely low cost processing [35].

As illustrated in figure 2.16, the materials selected for electron acceptor and electron donor are [6,6]-phenyl-C61-butyric acid methylester (PCBM) and P3HT, respectively. PCBM is a Buckminsterfullerene derivative (the 1996 Nobel Prize in Chemistry was awarded for the discovery of Buckminsterfullerene) that offers the advantage of having high electron affinity to produce efficient electron transfer and is also soluble in a liquid. P3HT is a conductive polymer (the 2000 Nobel Prize in Chemistry was awarded for its discovery) consisting of π-conjugated orbitals which are advantageous for photoactive devices.

The BHJ is deposited using the spray coating technique proposed in [35] where the materials are dissolved into a solvent and sprayed on to the substrate, offering significant cost reduction at the cost of surface roughness which can increase dark currents.

Each BHJ interface can be considered as a miniature *p–n* junction leading to an expanded Shockley equation (2.17) that defines the *I–V* relationship [18]:

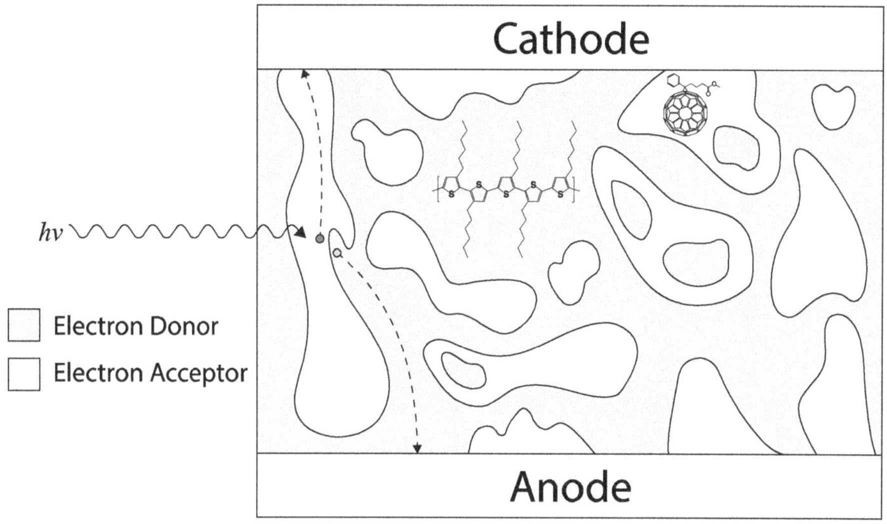

Figure 2.16. The bulk heterojunction concept made up of electron acceptor and electron donor including electron acceptor and electron donor materials, PCBM and P3HT, respectively.

$$I = I_0\left[\exp\left(\frac{qV}{n_{ID}k_BT} - 1\right)\right] \tag{2.37}$$

Notice that the single difference between the traditional and expanded Shockley equations is an extra variable in the denominator of the exponential term. The additional term is the so-called ideality factor n_{ID} that takes into account bulk morphology, and is illustrated in figure 2.17. Clearly, as $n \longrightarrow 0$ the diode reaches the I_0 saturation current when $V \longrightarrow 0$, which is advantageous since a lower bias voltage is required.

Organic semiconductors are typically vertical devices and therefore some insight into the device structure must be given. The substrate can be almost anything in organics including paper [36], plastic [37, 38] and glass [35]. The anode is generally made from transparent indium tin oxide (ITO) although there is a growing argument for using graphene due to the emergence of high efficiency devices with graphene anodes [39]. The next layers are the organic layers. In state-of-the-art OLED devices the organic layers are made up of (from bottom to top) a hole injection layer, a hole transport layer, an emissive layer, an electron transport layer, an electron injection layer followed by the cathode, which is generally aluminium since it is cheap and it is not necessary for it to be transparent. There are many devices that offer an increase in performance at the cost of increased complexity such as multiple photon emitters that are not covered here but are referred to in [40]. In BHJ OPDs the stack structure is significantly less complex; requiring only two electrodes, the BHJ and an optional interlayer; selected as P3HT in this thesis because it offers the highest bandwidth [41]. The interlayer is not covered here; however, it can have a profound effect on the performance of critical parameters of the device such as bandwidth; for a detailed

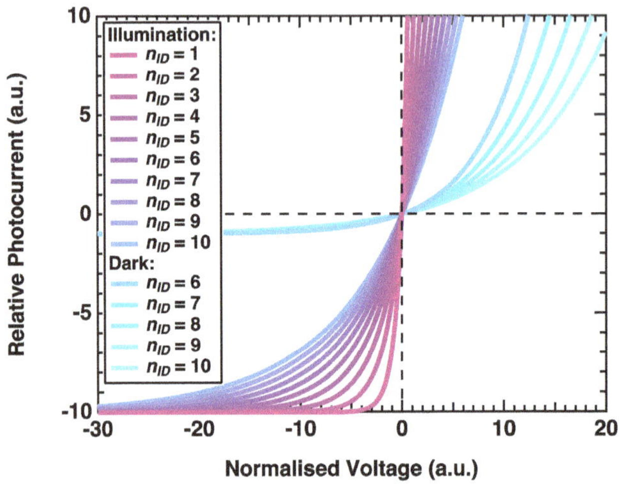

Figure 2.17. Shockley equation for an expanded *p–n* junction considering ideality factor *n*; the influence of *n* is illustrated clearly as with decreasing *n* the diode reaches the saturation current with less bias voltage, which is advantageous.

analysis, refer to [41]. As mentioned, the BHJ is an interpenetrated blend of P3HT: PCBM, which are extremely popular materials in BHJ devices due to their relatively high efficiency and solubility. The band-gap energy of P3HT:PCBM is ~2 eV, which is ideal for VLC applications as the cut-off wavelength is ~650 nm, which cuts a portion of the red wavelengths that would possibly be useful for wavelength division multiplexing (WDM). By introducing a further, low band-gap material into the BHJ blend such as poly[2,6-(4,4-bis-(2-ethylhexyl)-4Hcyclopenta[2,1-b;3,4-b′]dithiphene)-alt-4,7-(2,1,3-benzothiadiazole)] (PCPDTBT), the BHJ band gap can be reduced so the absorption spectrum extends into the NIR region and allows the absorption of such wavelengths. The working principles of P3HT:PCBM and similar BHJs are well covered in the literature and the reader is encouraged to refer to [18, 21, 35] since no details are given here.

2.4 Summary

This chapter has summarised the key technological enablers for VLC systems, mainly focusing on the LEDs and PD used in experimental systems, and providing pathways to be able to model the devices from their theoretical relationships. Beginning with basic *p–n* junction theory and radiative recombination of holes and electronics, resulting in the emission of photons. This theory was then related to the Shockley diode equation that defines the operation of all diodes.

The bandwidth limitations introduced by LEDs are then introduced through the junction capacitance that is the major contributor that defines it. Next, the chapter discussed equivalent circuit models for LEDs and PDs for the benefit of modelling at circuit level. These primaries will be useful later when discussing electronic equalisers in chapter 4. The major noise contributors were discussed including thermal and shot noise, and related to the specific detectivity of a PD.

Organic semiconductors were introduced in the same context, with their operating principles also discussed. Instead of having a fixed conduction and valence band as in inorganic semiconductors, organics consider molecular orbitals that can be considered as dislocated clouds of possible states with specific densities where electrons and holes can possibly exist. The mobility of holes and electronics inside organics is substantially lower than that of inorganics and hence a higher bandwidth limit is imposed.

References

[1] Sze S M and Ng K K 2006 *Physics of Semiconductor Devices* (New York: Wiley)
[2] Selberherr S 2012 *Analysis and Simulation of Semiconductor Devices* (Berlin: Springer)
[3] Chuang S L 2012 *Physics of Photonic Devices* vol 80 (New York: Wiley)
[4] Brütting W 2012 *Physics of Organic Semiconductors* (Weinheim: Wiley)
[5] Pfeiffer M, Leo K, Zhou X, Huang J S, Hofmann M, Werner A and Blochwitz-Nimoth J 2003 Doped organic semiconductors: physics and application in light emitting diodes *Organ. Electron.* **4** 89–103
[6] Li Q 2011 *Self-Organized Organic Semiconductors: From Materials to Device Applications* (New York: Wiley)
[7] Köhler A and Bässler H 2015 *Electronic Processes in Organic Semiconductors: An Introduction* (New York: Wiley)
[8] Fred Schubert E, Cho J and Kim J K 2000 *Light-Emitting Diodes* (New York: Wiley)
[9] Shockley W 1949 The theory of *p–n* junctions in semiconductors and *p–n* junction transistors *Bell Syst. Tech. J.* **28** 435–89
[10] Träger F 2012 *Springer Handbook of lasers and Optics* (Berlin: Springer)
[11] Levinshtein M E, Rumyantsev S L and Shur M S 2001 *Properties of Advanced Semiconductor Materials: GaN, AIN, InN, BN, SiC, SiGe* (New York: Wiley)
[12] Fox M 2002 *Optical Properties of Solids* (Oxford: Oxford University Press)
[13] Haigh P A, Ghassemlooy Z, Le Minh H, Rajbhandari S, Arca F, Tedde S F, Hayden O and Papakonstantinou I 2012 Exploiting equalization techniques for improving data rates in organic optoelectronic devices for visible light communications *J. Lightwave Technol.* **30** 3081–8
[14] Arca F, Sramek M, Tedde S F, Lugli P and Hayden O 2013 Near-infrared organic photodiodes *IEEE J. Quantum Electron.* **49** 1016–25
[15] Richards P L 1994 Bolometers for infrared and millimeter waves *J. Appl. Phys.* **76** 1–24
[16] Mackowiak V, Peupelmann J, Ma Y and Gorges A 2015 *NEP—Noise Equivalent Power* (Exeter: Thorlabs, Inc.)
[17] Stefan I, Elgala H and Haas H 2012 Study of dimming and LED nonlinearity for ACO-OFDM based VLC systems *IEEE Wireless Communications and Networking Conf.* (IEEE) pp 990–4
[18] Tedde S F 2009 Design, fabrication and characterization of organic photodiodes for industrial and medical applications *Thesis* Walter Schottky Institut, Technische Universitat Munchen
[19] Das R and Harrop P 2011 *RFID Forecasts, Players and Opportunities 2011–2021* (Santa Clara, CA: IDTechEx)
[20] Tang C W and VanSlyke S A 1987 Organic electroluminescent diodes *Appl. Phys. Lett.* **51** 913–5

[21] Brabec C J, Sariciftci N S and Hummelen J C 2001 Plastic solar cells *Adv. Funct. Mater.* **11** 15–26

[22] Krebs F C, Espinosa N, Hösel M, Søndergaard R R and Jørgensen M 2014 25th anniversary article: rise to power—OPV-based solar parks *Adv. Mater.* **26** 29–39

[23] Myny K, van Veenendaal E, Gelinck G H, Genoe J, Dehaene W and Heremans P 2012 An 8-bit, 40-instructions-per-second organic microprocessor on plastic foil *IEEE J. Solid-State Circuits* **47** 284–91

[24] Chaves L W F and Decker C 2010 A survey on organic smart labels for the internet-of-things *2010 7th Int. Conf. Networked Sensing Systems (INSS)* (IEEE) pp 161–4

[25] Baeg K-J, Caironi M and Noh Y-Y 2013 Toward printed integrated circuits based on unipolar or ambipolar polymer semiconductors *Adv. Mater.* **25** 4210–44

[26] Clayden J, Greeves N, Warren S and Wothers P *Organic Chemistry* (Oxford: Oxford University Press)

[27] Atkins P and De Paula J 2010 *Physical Chemistry* (Oxford: Oxford University Press)

[28] Frenkel J 1931 On the transformation of light into heat in solids. I *Phys. Rev.* **37** 17

[29] Frenkel J A 1931 On the transformation of light into heat in solids. II *Phys. Rev.* **37** 1276

[30] Wannier G H 1937 The structure of electronic excitation levels in insulating crystals *Phys. Rev.* **52** 191

[31] Pope M and Swenberg C E 1999 *Electronic Processes in Organic Crystals and Polymers* (Oxford: Oxford University Press)

[32] Roichman Y and Tessler N 2002 Generalized Einstein relation for disordered semiconductors-implications for device performance *Appl. Phys. Lett.* **80** 1948–50

[33] Kaake L G, Barbara P F and Zhu X-Y 2010 Intrinsic charge trapping in organic and polymeric semiconductors: a physical chemistry perspective *J. Phys. Chem. Lett.* **1** 628–35

[34] Miller A and Abrahams E 1960 Impurity conduction at low concentrations *Phys. Rev.* **120** 745

[35] Tedde S F, Kern J, Sterzl T, Furst J, Lugli P and Hayden O 2009 Fully spray coated organic photodiodes *Nano Lett.* **9** 980–3

[36] Eder F, Klauk H, Halik M, Zschieschang U, Schmid G and Dehm C 2004 Organic electronics on paper *Appl. Phys. Lett.* **84** 2673–5

[37] Someya T 2010 Flexible electronics: tiny lamps to illuminate the body *Nat. Mater.* **9** 879

[38] Wang Z B, Helander M G, Qiu J, Puzzo D P, Greiner M T, Hudson Z M, Wang S, Liu Z W and Lu Z H 2011 Unlocking the full potential of organic light-emitting diodes on flexible plastic *Nat. Photonics* **5** 753

[39] Han T-H, Lee Y, Choi M-R, Woo S-H, Bae S-H, Hong B H, Ahn J-H and Lee T-W 2012 Extremely efficient flexible organic light-emitting diodes with modified graphene anode *Nat. Photonics* **6** 105

[40] Sasabe H, Takamatsu J-i, Motoyama T, Watanabe S, Wagenblast G, Langer N, Molt O, Fuchs E, Lennartz C and Kido J 2010 High-efficiency blue and white organic light-emitting devices incorporating a blue iridium carbene complex *Adv. Mater.* **22** 5003–7

[41] Arca F, Tedde S F, Sramek M, Rauh J, Lugli P and Hayden O 2013 Interface trap states in organic photodiodes *Sci. Rep.* **3** 1324

Part I

Visible light for data communication

IOP Publishing

Visible Light
Data communications and applications
Paul Anthony Haigh

Chapter 3

High-speed circuits and channel modelling

3.1 Introduction

In this chapter, high-speed circuits and channel models will be discussed, which are two separate self-contained research topics. Here, they are combined, since the bandwidths achievable in the channel are often not limited by the environmental characteristics such as room dimensions and multi-path limitations, but by the slow frequency response of the LEDs themselves. Therefore, to achieve predicted channel capacities, the system must be co-designed with these factors in mind. The first step in the process is to equalise the bandwidth of the LED, to extend it, mitigating its data rate limitations before modelling the channel to predict the multi-path response. Hence, in this chapter the two topics are presented together.

3.2 Circuits for high-speed VLC systems

Circuits form an integral, if often overlooked, part of the design of VLC systems. Considering the wide range of LEDs and their different types (i.e. WPLED, RAGB, PLED, μLED, etc) and the variations from vendor-to-vendor, it is necessary to consider bespoke circuit designs for any LED used for VLC. The same is true at the receiver side where the PD used has its own characteristics that require bespoke design.

There have been a number of circuit design techniques reported in the recent literature [1–9]; however, there has not been one distinct and holistic approach, with many wide-ranging approaches proposed. For instance, it is widely known that LEDs are current sinks and should be driven with a current-led approach. However, the vast majority of the literature elects to drive the LEDs using a bias tee, which will be discussed later, to bias and modulate the device in a voltage-led manner, which means that optimal performance cannot be obtained, and results in low modulation depths. Other approaches, such as using a fast NAND gate with an open-collector transistor based output have been proposed that increase modulation depths [10],

doi:10.1088/978-0-7503-1680-4ch3

but force a limit on modulation formats to pulse-based schemes due to the on-off nature of the digital logic used to impress the data onto the dc current.

Depending on the application, the driving circuit must also be adjusted. For instance, if one requires a high-speed driving circuit, amplitude equalising properties must be included in the circuit design. On the other hand, if one requires a dimmable circuit, then a pulse width modulation (PWM) approach is often adopted, along with the relevant high-speed drivers [7, 11, 12]. In this section, a number of innovations in high-speed circuits are discussed.

3.2.1 Multiple-resonant equalisation

Since LEDs have low modulation bandwidths in the order of several MHz [13], one way to improve their bandwidth is to use electronic components in the driving circuits that extend the bandwidth through interaction in the frequency domain. One such example of a pre-equaliser was demonstrated experimentally in [2] through a process called multiple-resonant equalisation. In this method, the driver circuit consists of several resistor-inductor capacitor (RLC) filters, each with a different peak output frequency, which when combined with each other and an LED, extend the bandwidth via a net sum of all of the frequency responses designed. The concept behind this figure is illustrated in figure 3.1, where 16 LEDs are considered in a matrix panel.

The resonant circuit is effectively a bias tee with a series inductance placed before the LED. The capacitor blocks the dc from flowing into the buffer output and the inductor L_{dc} prevents the ac from flowing into the dc source, thus forcing them to mix and flow to the LED. The circuit diagram is outlined in figure 3.2.

In [2], a total of eight resonant circuits are considered for the 16 LEDs with the parameters in table 3.1. The LEDs are set in a 4×4 matrix with a 60 mm pitch, with

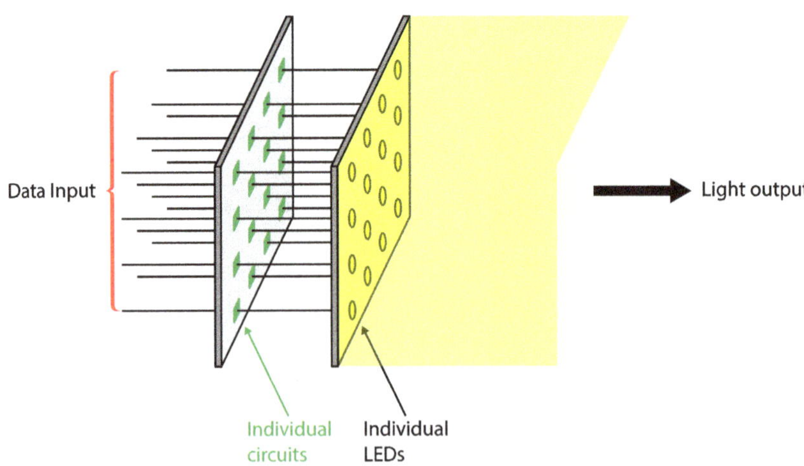

Figure 3.1. A conceptual illustration of the multiple-resonant equalisation scheme. Each LED has its own driver that equalises a portion of the magnitude response, resulting in an aggregately larger bandwidth at the receiver.

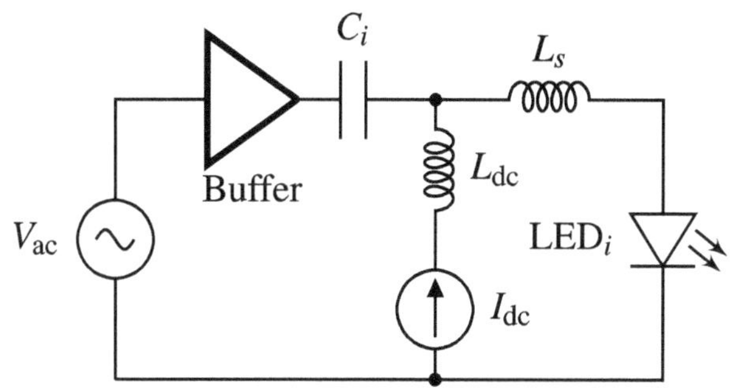

Figure 3.2. The multiple-resonant circuit diagram, including bias tee and LED.

Table 3.1. Table of components used for each of the LEDs in the matrix, adopted from [2].

LED Index	Capacitor	Inductor	a (dB)
1	10 μF	330 nH	−13
2	4.7 nF	330 nH	−13
3	820 pF	330 nH	−6
4	680 pF	330 nH	−6
5	560 pF	330 nH	−6
6–8	330 pF	330 nH	0
9–11	220 pF	330 nH	0
12–16	150 pF	330 nH	0

a total 1.5 W power output. Each LED is driven at 200 mA and the received illumination was fixed at 400 lx. The higher frequency resonant peaks are repeated several times to add magnitude to those frequency components, hence a number of capacitors are repeated across the 16 LEDs. An attenuation coefficient a is also introduced that balances the relative magnitudes of the peaks to ensure a flatness of the LED frequency response.

The driving circuit consists mainly of a resonant RLC circuit tuned to a specific frequency, which is used to both equalise the LED, modulate and bias it simultaneously. The magnitude response $|H_i(\omega)|$ of the ith individual circuit is given by [14]:

$$|H_i(\omega)| = \frac{\omega R C_i}{\sqrt{(1 - \omega^2 L_s C_i)^2 + (\omega R C_i)^2}} \tag{3.1}$$

where $\omega = 2\pi f$ is the angular frequency, L_s is the total series inductance (including that in the LED), C_i is the resonant capacitor and R is the internal resistance of the LED. The resonant frequency of the ith circuit is given by $f_i = (2\pi\sqrt{L_s C_i})^{-1}$. The total value of L_s is obtained by calculation of the gradient of $1/\sqrt{C_i}$ and was estimated to be 330 nH from [2]. The overall bandwidth of the multiple-resonant circuit is therefore controlled by adjusting the component values to control the resonant frequency of the individual drivers, which in turn alters the magnitude

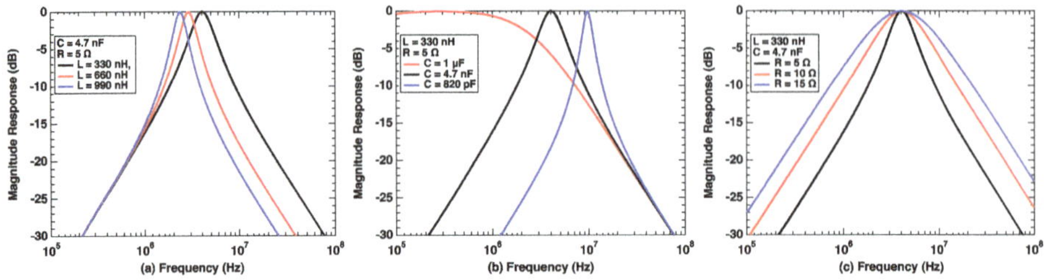

Figure 3.3. The impact of changing individual components in the multiple-resonant circuit design; changing (a) the inductance results in a slight upward frequency shift and broadening of the spectrum, (b) a sharpening of the spectrum with an opposite frequency shift and (c) a broadening of it without a shift.

responses of the LEDs. The final superposed magnitude response $F(\omega)$ is given by [2]:

$$F(\omega) = \sum_{i=0}^{N_{\text{LEDs}}} H_i(\omega)a_i \tag{3.2}$$

where N_{LEDs} is the number of LEDs in the circuit and a_i is a scaling factor that controls the amplitude of the ith LED magnitude response. To illustrate the impact of the individual components, figures 3.3(a)–(c) show the result of varying the inductor, capacitor and resistor, respectively. Increasing the inductance causes a shift to lower frequencies and a narrowing of the passband, while increasing the capacitance has the opposite effect, widening the passband with increasing capacitance, both as expected from electronics fundamentals. Finally, increasing the resistance results in no frequency shift but a widening of the passband. These are all important considerations when considering which values to set in the equalisation circuits.

In figure 3.4 the normalised frequency peaks of the equaliser designed in [2] without modification of the amplitude are shown to give an idea of the spacing of the resonant frequency peaks for each individual LED driver circuit. Simultaneously, in figure 3.5, the equaliser peaks with the attenuation coefficients applied are shown, as well as the resulting aggregated magnitude response. Certainly, it is worth noting that although convolving this equaliser will improve the LED bandwidth (refer to [2], not shown here), it introduces a band-pass filter (BPF) shape due to missing energy as highlighted. This is caused by the introduction of numerous capacitors in the transmitter circuity that block low frequencies and therefore introduce the negatively impacting baseline wander (BLW) phenomena, which makes hard threshold detection difficult [15].

3.2.2 Artificial transmission line synthesis

As mentioned, the plate capacitance of the LED is the limiting factor to its modulation bandwidth. While the previous method manages to ameliorate this consideration it also requires numerous coupling capacitors that introduce a heavy BLW effect. To overcome this limitation, the authors of [16] proposed a new method

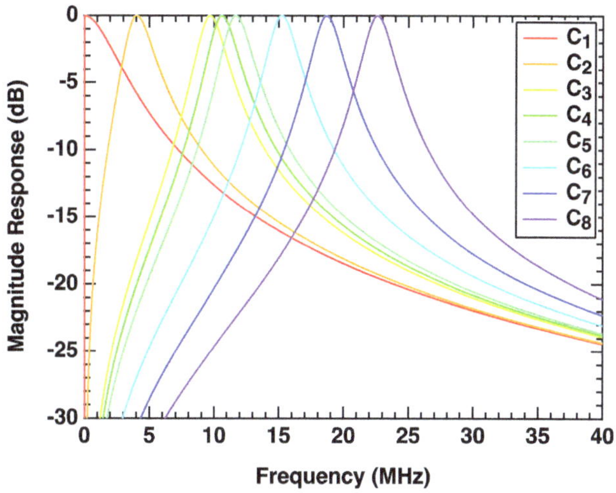

Figure 3.4. The resonant peaks of the equaliser designed in [2] before amplitude correction.

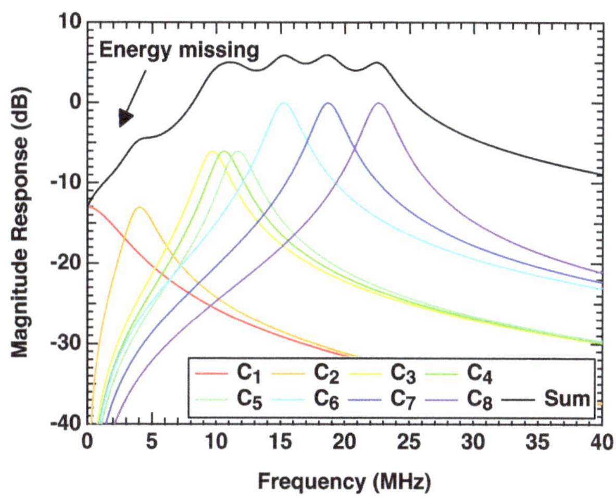

Figure 3.5. The resonant peaks of the equaliser designed in [2] after amplitude correction.

to improve the LED bandwidth that diffuses the equivalent plate capacitance into pseudo-artificial transmission lines (p-ATLs) that offer an inductor-like response. This technique is based on distributed transmission line synthesis, a technique that has been adopted widely in ultra-wide-band amplifier design [17–21]. A similar approach was reported in [22], where residual charge carriers in the LED device were drawn out into the circuit via a push-pull circuit. The proposed p-ATL approach is advantageous because it is compatible with any bespoke driver circuit based on individual characteristics of any given LED, and can be used in combination with existing bandwidth extension schemes, which may lead to further improvement in available modulation bandwidth.

The simplified LED circuit previously outlined in chapter 2 is illustrated once more, where C_d is the diffusion capacitance, R_d is the small signal dynamic resistance and R_s is the ohmic contact resistance of the LED. The input impedance of this model can be obtained as a function of the RC time constant [23] by circuit analysis as [16] (figure 3.6):

$$Z_{in} = R_s + \frac{R_d}{1 + j\omega R_d C_d} \tag{3.3}$$

Given a fixed bias, LEDs generally exhibit a first order LPF response $f_c = (2\pi C_d R_d)^{-1}$, and clearly it is C_d that must be incorporated into the p-ATL via a series inductance. To distribute the capacitance across a p-ATL, a parallel inductor L_d is introduced to the equivalent circuit with a load resistor as shown in figure 3.7. The value of the inductor is selected based on the image impedance (Z_o) relation for a lossless p-ATL, given as [16]:

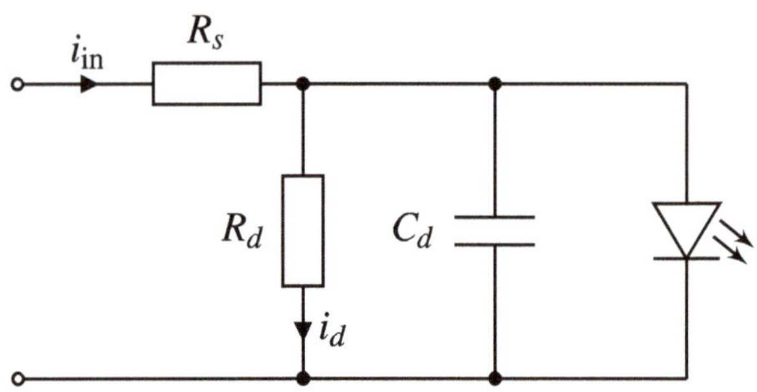

Figure 3.6. The simplified LED equivalent circuit.

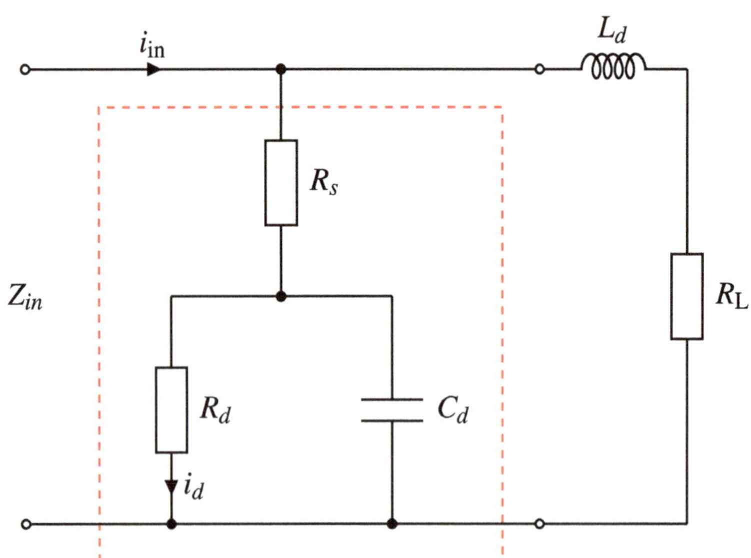

Figure 3.7. Simplified LED equivalent circuit with distributed input.

$$Z_o = \sqrt{\frac{L_d}{C_d}\left[1 - \frac{\omega^2 L_d C_d}{4}\right]} \tag{3.4}$$

and the cut-off frequency of the p-ATL, $f_{c\text{-p-ATL}}$, given by [16]:

$$f_{c\text{-p-ATL}} = \frac{\omega_c}{2\pi} = [\pi\sqrt{L_d C_d}]^{-1} \tag{3.5}$$

where ω_c is the angular cut-off frequency. Using (3.4), L_d is selected as the product of C_d and the square of the dc value of Z_o [24]. The terminating resistor is selected as the dc value of Z_o to achieve a broadband match. It may be observed from figure 3.7 that a current divider circuit is formed between the LED and the sum of R_L and the reactance of L_d, which in turn reduces the LED biasing current. Both R_d and C_d are dependent on the bias, and hence, the p-ATL must therefore be designed at the parameters corresponding to the desired effective LED drive current. Another important factor to consider is that the parallel combination of $(R_d + R_s)$ and R_L also impacts Z_o, such that R_L does not provide an image impedance match even at low frequencies.

The input impedance of the LED with distributed input Z_{in}^* shown in figure 3.7 is derived as [16]:

$$Z_{in}^* = \left[R_s + \left[\frac{R_d}{1 + j\omega R_d C_d}\right]\right] // (j\omega L_d + R_L) \tag{3.6}$$

which, through expansion, becomes [16]:

$$Z_{in}^* = \frac{R_L(R_s + R_d) + j\omega(R_s R_d R_L C_d + L_d(R_s + R_d)) - R_s\omega^2 R_d C_d L_d}{R_L + R_s + R_d + j\omega(L_d + R_s R_d C_d + R_d R_L C_d) - \omega^2 R_d C_d L_d} \tag{3.7}$$

Clearly, from (3.7) it is possible to deduce that the frequency response and cut-off behaviour of the LED is no longer uniquely dependent on the RC time constant, and hence, through appropriate selection of Z_o a significant improvement in bandwidth can be obtained. Furthermore, at dc $Z_{in} = (R_s + R_d) = R_L$ is clearly lower than Z_{in} and hence while there is a gain in bandwidth, there is also an associated loss in impedance magnitude. Sacrificing the impedance magnitude for a wider bandwidth means that the optical power output and light intensity are reduced, which may not be an advantageous characteristic. Hence, a trade-off between bandwidth extension and output power must be reached.

To illustrate the performance of the p-ATL technique, figure 3.8 highlights the input impedance of an LED in two separate cases; i.e. $Z_o = Z_{in}$, and $Z_o = Z_{in}/2$ with the assumption of equal effective drive current. The LED equivalent model parameters were extracted from [25] and are presented in table 3.2. The 2 mA drive current I_d was used in figure 3.8.

As reported in [25], the LED equivalent circuit has a modulation bandwidth of 18.7 MHz. When distributed transmission line impedance is selected for $Z_o = Z_{in}$, it can be inferred that a threefold improvement in bandwidth can be obtained, while

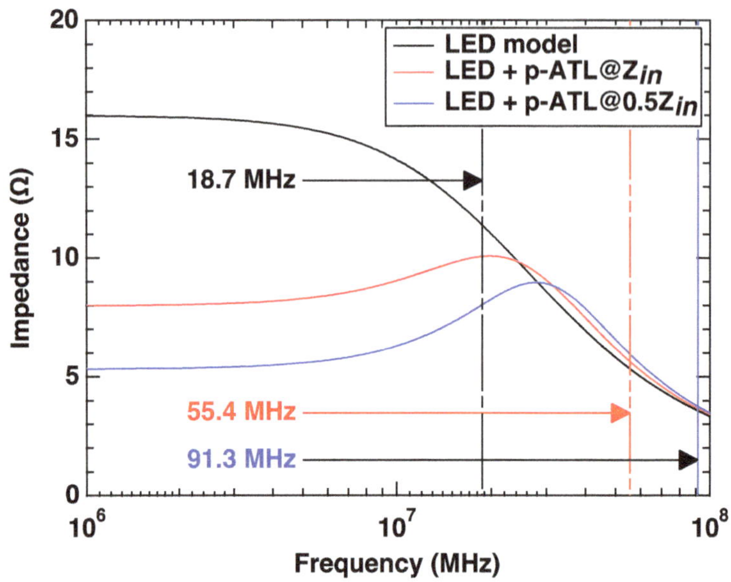

Figure 3.8. Impedance response of the p-ATL approach to LED equalisation; the bandwidth is extended by adding an inductor and load resistor in parallel with the LED at the cost of power due to resistive loading and resonant peaking.

Table 3.2. Component values used in [16] for the LED equivalent circuit at different bias currents.

I_b (mA)	R_s (Ω)	R_d (Ω)	C_d (nF)
1	1.6	28.4	0.342
2	1.6	14.4	0.59
5	1.6	6.0	1.33
10	1.6	3.0	2.67

the input impedance reduces by ~50%, in linear agreement with (3.7). For the second case where $Z_o = Z_{in}/2$, there is an additional further improvement in bandwidth of 65%, which translates to a ~400% overall improvement in bandwidth when compared to the LED performance. However, this comes at the cost of a 66.7% overall loss in impedance magnitude, significantly reducing optical power intensity output.

3.2.3 Negative impedance conversion

Another circuit design that is of considerable interest within the VLC community is the realisation of negative impedance through active circuit design. Negative capacitance as an overall concept has been used in numerous electronic applications including amplifier and filter design [26–28]. To realise negative impedance, it is necessary to deploy a negative impedance converter (NIC), which is effectively a two

I_b (mA)	R_s (Ω)	R_d (Ω)	C_d (nF)
1	1.6	28.4	0.342
2	1.6	14.4	0.59
5	1.6	6.0	1.33
10	1.6	3.0	2.67

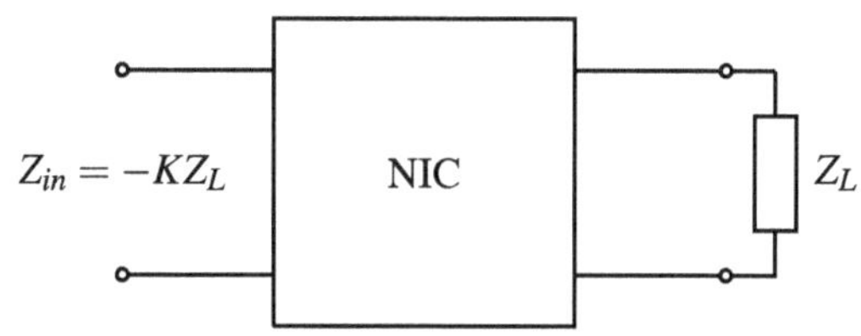

Figure 3.9. The negative impedance conversion two port concept. The load impedance Z_L is projected at the input ports as $-Z_L$.

port network that, when terminated by a load impedance at one port, presents a negative impedance at the other. This concept of negative impedance conversion is simple and a general block diagram is highlighted in figure 3.9.

In [29], the first report of negative capacitance was introduced in parallel with the bandwidth-limiting LED diffusion capacitance to aggregately reduce the overall capacitance and hence, improve the LED bandwidth. The input impedance of the port illustrated in figure 3.9 can be given as $Z_{in} = -KZ_L$, where the ideality factor is given by K and the load impedance is given by Z_L. When $K = 1$, the NIC can be considered ideal. A NIC is generally designed to either generate a negative version of a reactive element, be it resistor, capacitor or inductor and the type generated depends on the load connected to the load impedance. In the case of VLC, due to the fact that the equivalent capacitance of the LEDs is the limiting factor, the most useful component to negate would be the capacitor in parallel, as illustrated in the conceptual circuit diagram in figure 3.10, and the negative resistance $-R_c$ is included for completeness.

The simplest form of a current inversion NIC circuit is illustrated in figure 3.11, and it makes intelligent use of a common collector stage (Q1) along with a common base stage (Q2). Feedback is established through connection of the output and input of the common collector stage through the low impedance Q2 stage. Through this process, the load impedance of C_c placed between the two emitters of Q1 and Q2 is negated at the input and is manifested as a negative capacitance. If $K = 1$, this value will be $-C_c$.

The input impedance of the NIC is described in [29] through small signal analysis and the assumption that ideal components were modelled using simple dc-circuit models, hence internal parasitics are omitted. The input impedance is given by [29]:

$$Z_{in} \approx \frac{g_{m2}(1 + g_{m1}r_{\pi 1}) + k\omega C_c(1 + g_{m1}r_{\pi 1} + g_{m2}r_{\pi 1})}{j\omega C_c R_{\pi 1} g_{m1} g_{n2}} \tag{3.8}$$

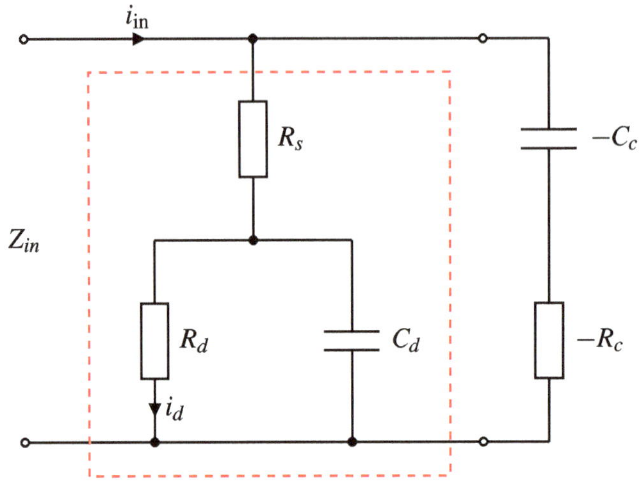

Figure 3.10. Simplified equivalent circuit of the NIC concept where the components are projected in parallel with the LED.

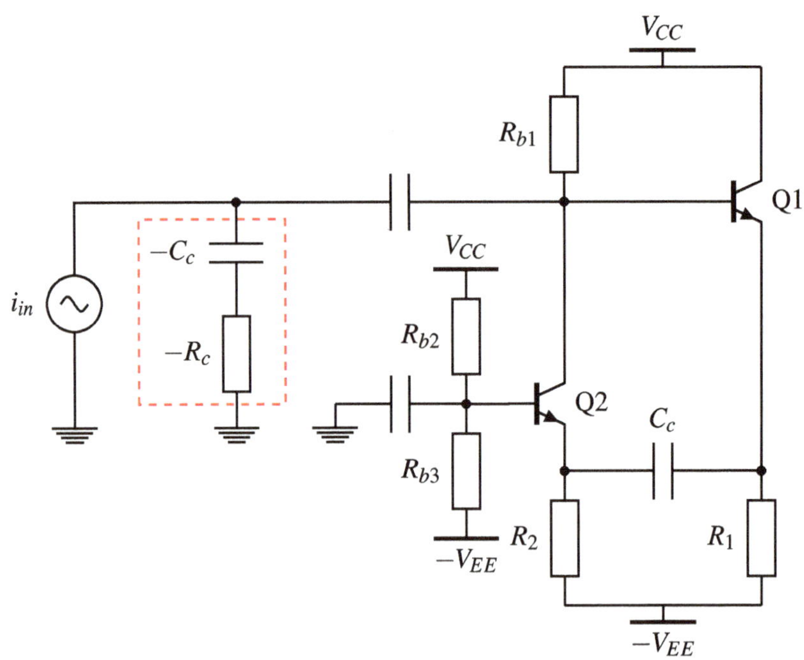

Figure 3.11. The circuit developed for the experimental implementation of the current inversion NIC [29].

where g_{mi} is the transconductance and r_{pi} is the internal resistance of the ith transistor. The resistance is given by [29]:

$$R_c \approx -\frac{g_{m1} + g_{m2}}{g_{m1} g_{m1} g_{m2}} \tag{3.9}$$

and finally the reactance is given as follows [29]:

$$X_c \approx \frac{1}{\omega C_c} \tag{3.10}$$

Clearly, from the above, it is possible to infer that the input impedance of the NIC is a combination of the negative capacitance related to the terminating capacitor C_c and the negative resistance component which arises from the internal resistance of the transistors Q1 and Q2 and their biasing conditions. The value of the negative resistance is important because it influences the value of K [29]. Thus, the negative values of capacitance and resistance generated can be modelled as separate components, as highlighted in figure 3.11.

Recalling the circuit model illustrated in the dashed box of figure 3.6, the input impedance is given in (3.3). By amelioration of the capacitance C_d through introduction of $-C_c$, bandwidth extension is possible, since the aggregated bandwidth f_c relationship is defined as [29]:

$$f_c = \frac{1}{2\pi R_d(C_d - C_c)} \tag{3.11}$$

where if $C_d \equiv C_c$, the bandwidth would be increased indefinitely, which could obviously never be realised experimentally due to the internal resistance of the transistors and other parasitics.

By circuit analysis, the overall input impedance can be obtained as follows [29]:

$$Z_{in} = \frac{(R_d R_s C_s s + R_d + R_s)(R_c C_c s + 1)}{(R_d C_d C_c (R_s - R_c))s^2 + (R_d(C_d - C_c) - C_c(R_s - R_c))s + 1} \tag{3.12}$$

where the dc resistance is given by $R_s + R_d$ and the transfer function clearly contains two real-valued zeros, i.e. $w_{z1} = (R_s + R_d)(R_d R_s C_d)^{-1}$ and $w_{z2} = (R_c C_c)^{-1}$. When $R_s \neq |R_c|$, there is a complex pole, and for the condition when $R_s = |R_c|$, there is just one real-valued pole, refer to [29–31] for more information.

It follows that there are several key notes that form the basis of this equaliser. Firstly, the LED bandwidth is no longer simply dependent on C_d and R_d as previously. Now, due to the intervention of the NIC circuit, it also depends on R_c, which can be tailored based to the individual LED, ensuring that $R_s = |R_c|$ and as such, the overall achieved bandwidth is controlled by the value of R_d specifically, alongside the summed capacitances of C_d and C_c. Furthermore, the equivalent resistance of the LED is not influenced at all by the NIC, and hence, there is no loss in LED optical power output, and resistive loading is avoided. Importantly, and in contradiction to the previously outlined techniques, there is therefore not a bandwidth-power trade-off that costs optical power. The impedance of the LED model compared with that of three equalisers with varying negative capacitances can be seen in figure 3.12, where it can be inferred that increasing the negative capacitance results in a higher bandwidth, but also results in some peaking if the negative capacitance is increased excessively.

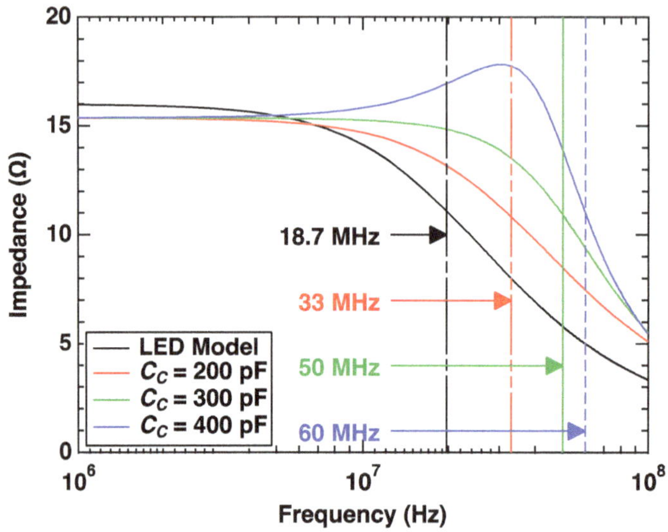

Figure 3.12. The impedance response of the LED when driven by the negative impedance circuit, showing a bandwidth extension up to 60 MHz [29].

3.2.4 Other solutions

There have been numerous solutions proposed for VLC to extend the bandwidth of LEDs over the years [32–38]. The vast majority of which use resistive loading and hence, introduce a trade-off between optical power and bandwidth. Potentially, this is not an attractive prospect as following Shannon's famous capacity equation [39]:

$$C = B\log_2(1 + \text{SNR}) \tag{3.13}$$

it can be inferred that trading SNR for bandwidth and vice-versa may not be be the optimal solution, when resistive loading can be avoided, as in the case of the NIC. There simply isn't sufficient space to detail each equalisation technique proposed in the literature, and hence this sub-section is concluded. The next important topic to be covered is that of channel models, which can also cause band-limitations or other negative impacts on the link.

3.3 Channel models

The channel in a VLC system is a free-space channel that depends on several parameters such as whether the link is (i) line-of-sight (LOS), where the signal simply falls away with the square of the distance and depends heavily on the angle of emission or absorption, or (ii) non-line-of-sight (NLOS), where the dimensions of the indoor environment, the materials of the walls and the number of reflections that the light undergoes between transmitter and receiver are all of key importance. In LOS configurations, there are two further classifications, directed, where the transmitter and receiver are directly aligned, and non-directed, where the transmitter emits a wide-angled spot-light and the receiver exists somewhere within its field-of-view (FoV).

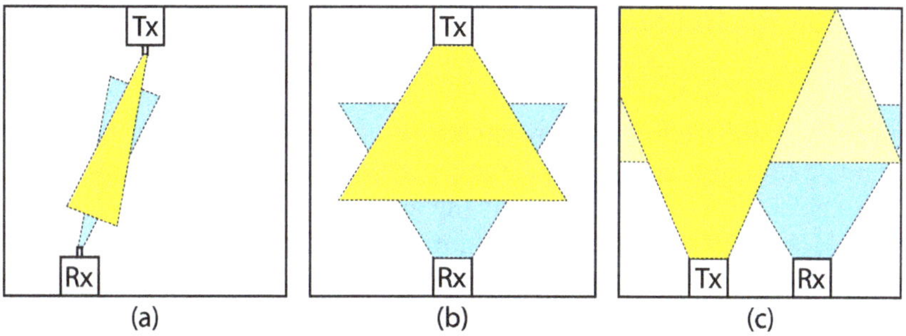

Figure 3.13. Possible scenarios of VLC systems including (a) directed LOS, (b) non-directed LOS and (c) diffuse. The most common application for generic VLC systems is non-directed LOS; however, most high-speed links demonstrated in the literature utilise the directed LOS as more power lands at the receiver.

In NLOS mode, both the transmitter and receiver can be located anywhere, provided there is a link consisting of sufficient power present via a multi-path propagation. These are highlighted in figures 3.13(a)–(c), which show (a) directed LOS, (b) non-directed LOS and (c) NLOS. The discussions provided in this book mainly focus on non-directed LOS links which are typically used in VLC.

The generic VLC channel model in terms of the dc gain may be given as a continuous integral as follows [40, 41]:

$$H(0) = \int_{-\infty}^{\infty} h(t)\mathrm{d}t \tag{3.14}$$

where $h(t)$ is the channel impulse response. Furthermore, the dc channel gain can be related to the average received optical power P_r following [42]:

$$P_r = P_t H(0) \tag{3.15}$$

where P_t is the transmitted power. In a static channel, the impulse response can be split into two components, the LOS and the NLOS. The total receiver power can be therefore defined as [43]:

$$P_r = P_t H_d(0) + \int P_t H_{ref}(0) \tag{3.16}$$

where H_d and H_{ref} are the LOS and NLOS components, respectively. The received SNR is given as the ratio of the received signal power to the noise power N_0 [44]:

$$\mathrm{SNR} = \frac{\Re^2 P_t^2 H(0)^2}{B N_0} \tag{3.17}$$

where $\Re$ is the responsivity of the PD, B is the signal bandwidth and N_0 is the noise spectral density in W/Hz. For a static channel, the SNR is dominated by the relationship between P_t (or P_r) and B, since $\Re$, $H(0)$ and N_0 are fixed.

In indoor applications, generally a static channel is commonplace in the literature, which means there is no free movement of objects or people and the environment is controlled. Moreover, within such an environment, a high SNR is typically available [45] due to the high optical power and highly sensitive PIN diodes generally used. For instance, such a channel is analysed in [46], which looked at different impulse responses in the indoor environment based on different room configurations and multi-path propagation effects via Lambertian reflectors. Different available link bandwidths were derived based upon the different room configurations and it was found that for the given simulation scenarios, bandwidths up to ~34 MHz could be achieved, presuming that the optoelectronic devices were free from bandwidth limitations. The reason that the bandwidth of the channel was limited was due to the delay spread induced by the delay between the arrival of signals from the LOS component and the component that undergoes a high number of reflections, which causes a severe quantity of inter-symbol interference (ISI). This work was later updated [47] to analyse the impact of different materials such as plaster and plastic on the multi-path propagation, light diffusion and delay spread (and consequently, ISI).

The generalised Lambertian radiation pattern is illustrated in figure 3.14. The radiation pattern of Lambertian emission $R(\phi)$ is given by [46]:

$$R(\phi) = \frac{m+1}{2\pi} P_t \cos^m(\phi) \quad \text{for } \phi \in [-\pi/2, \pi/2] \tag{3.18}$$

where m is the Lambertian order and ϕ is the angle of emission. The Lambertian order m is an important quantity because it describes the overall shape of the emitted radiation. When $m = 1$, the emission is omnidirectional away from the source along the emission plane, while when $m \longrightarrow \infty$ the beam becomes more directional, as can be seen in figure 3.14. The Lambertian order m is defined in turn by the LED semi-angle at a half illuminance $\Phi_{1/2}$.

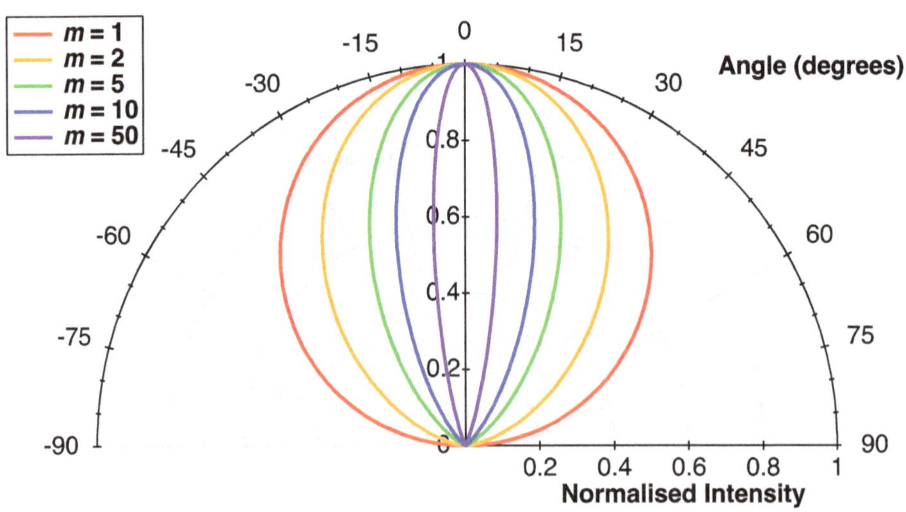

Figure 3.14. Radiation emission patterns for a variety of Lambertian orders, m. As $m \longrightarrow \infty$, the beam becomes more directional. Directed optics such as LDs typically have high values of m, while LEDs have lower values and larger optical footprints.

In turn, m is given by [41, 48]:

$$m = \frac{\ln(2)}{\ln\left[\cos\left(\Phi_{1/2}\right)\right]} \tag{3.19}$$

At the receiver, the channel model must take advantage of the above emission pattern which is common for LEDs, and must also take into account the distance and physical aspects of the PD used, such as its photoactive area, FoV and the distance is placed away from the transmitter. One must also consider that the received optical power may undergo one or more reflections on its path to the receiver. An illustration of this is shown in figure 3.15. The two channel model components considering all of these parameters are given by [49]:

$$H_d(0) = \begin{cases} R(\phi) A_{PD} \cos(\theta) & \text{for } 0 \leqslant \theta \leqslant \theta_{\text{FoV}} \\ 0 & \text{for } \theta > \theta_{\text{FoV}} \end{cases} \tag{3.20}$$

and

$$H_{ref}(0) = \begin{cases} \rho R(\phi_r) \dfrac{A_{PD} A_w \cos\left(\theta_r\right) \cos\left(\alpha\right) \cos\left(\beta\right)}{\pi (d_1 d_2)^2} & \text{for } 0 \leqslant \theta \leqslant \theta_{\text{FoV}} \\ 0 & \text{for } \theta > \theta_{\text{FoV}} \end{cases} \tag{3.21}$$

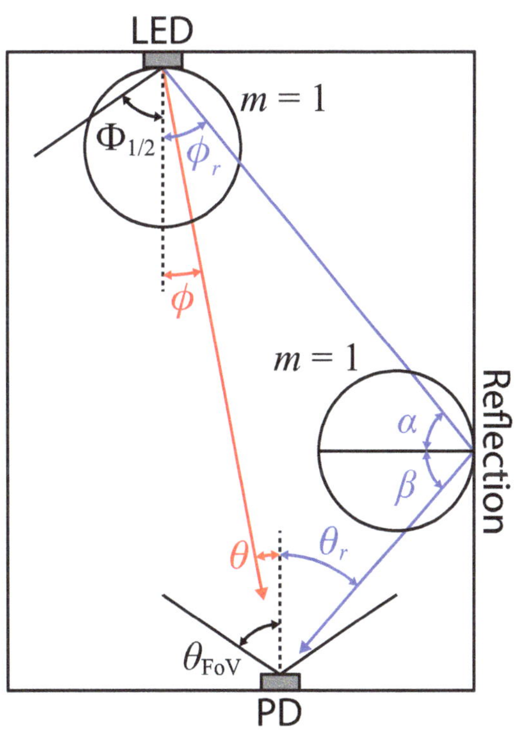

Figure 3.15. A typical scenario that considers reflections in a broadcasting network. The PD has a wide FoV and can collect light from the LOS path but also reflective paths, leading to inter-symbol interference and bandwidth limitations that must be modelled.

respectively, where θ_r is the angle of emission for the reflected path, A_{PD} is the area of the PD, θ (LOS) and θ_r NLOS are the angles of incidence to the detector and θFoV is its FoV. In the NLOS component, there are additional components which are ρ, the reflectance coefficient, α and β the angles of incidence and irradiance from the point of reflection, and d_1 and d_2, the distances between the transmitter and the reflection point, and the distance between the reflection point and the receiver, respectively. Finally, A_w is the area of reflectance considered. In some cases, an optical filter and concentrator are added as coefficients in the above equations, however, they are omitted here to maintain simplicity.

Assuming any points of reflection behave as Lambertian reflectors, i.e. that they absorb and re-emit some fraction of the optical power as a secondary Lambertian emitter, the multi-path impulse response component after the rth reflection can be generalised using the above equation, solving for each specific case in a ray-tracing manner.

Another consideration of the multi-path environment is the impact on the time-of-arrival that the reflections have on the signal. Until now, (3.14)–(3.21) have not considered any temporal effects and therefore the channel model has been defined by its dc characteristics. Clearly, due to the nature of the multi-path environment where path lengths are not equal, it is necessary to consider the impact of this on the communications performance. Two key performance indicators commonly used in wireless RF communications are the root mean square (RMS) delay spread and the mean excess delay. The formal definitions of these two are well defined in the literature and can be found in numerous texts [49–52]. The RMS delay spread is given by [49]:

$$\tau_{\mathrm{RMS}} = \sqrt{\frac{\int_{-\infty}^{\infty} (\tau - \tau_0)^2 h^2(t)\mathrm{d}t}{\int_{-\infty}^{\infty} h(t)\mathrm{d}t}} \tag{3.22}$$

where τ is the propagation time and τ_0 is the mean excess delay given by [49]:

$$\tau_0 = \frac{\int_{-\infty}^{\infty} \tau h^2(t)\mathrm{d}t}{\int_{-\infty}^{\infty} h^2(t)\mathrm{d}t} \tag{3.23}$$

where $h(t)$ is the channel response in the time domain, given as the inverse Fourier transform of the frequency domain response.

It is clear from (3.22) that τ_{RMS} is influenced heavily by the environment. If there are numerous reflections and/or long path lengths, then the RMS delay spread will increase, resulting in lower bandwidths and higher ISI due to the smearing of energy between symbols. On the other hand, when there are few reflections or the path lengths are similar, smaller values of τ_0 are obtained, which result in an improved available bandwidth. The channel coherence bandwidth is the range of frequencies that the channel response can be considered flat, and is expressed by [53]:

$$B_{c,50\%} = \frac{1}{5\tau_0} \tag{3.24}$$

where the 50% indicates the de-correlation percentage. Thus, the maximum achievable data rate R_b using binary signalling that can be recovered from the channel is given by [54]:

$$R_b \leqslant \frac{1}{10\tau_0} \tag{3.25}$$

The 3 dB bandwidth of the of the channel is simply found as the point in which the channel magnitude response drops to half, i.e. [41, 55]:

$$|H(f_{3dB})|^2 = \frac{1}{2}|H(0)|^2 \tag{3.26}$$

The time-domain interpretation of $h(t)$ considers all the reflections from the multi-path channel, where N_r is the number of reflections [55, 56]:

$$h(t) = h_{LOS}(t) + \sum_{n}^{N_r} h_{ref}^{n}(t) \tag{3.27}$$

Different environments have a different response because they have different geometric sizes and moreover, rooms with different materials that possess different reflectivities also cause multi-path contributions to vary [46, 47]. The impact of high-order reflections that reduce bandwidth caused the emergence of research in angle-diverse receivers, which are capable of establishing LOS paths (or paths with a low-order of reflections). One such system was demonstrated in [57], where a receiver that consists of six detectors, each facing a unique direction outwards separated by an angle of 30° and covering a FoV of 60° is presented. A seventh detector faces vertically upwards (in case of a LOS link) and the overall concept is illustrated in figure 3.16. After recovery of the optical signal, each detector generates a photocurrent that is converted to a voltage in the usual way by a TIA. The signal power is measured for each incoming link and fed to a microcontroller, which compares the received signal strength. Whichever PD has the strongest signal is selected and the information from that link is then used for post-processing and demodulation. Using this method, a low (10^{-6}) bit-error rate (BER) can be achieved regardless of the

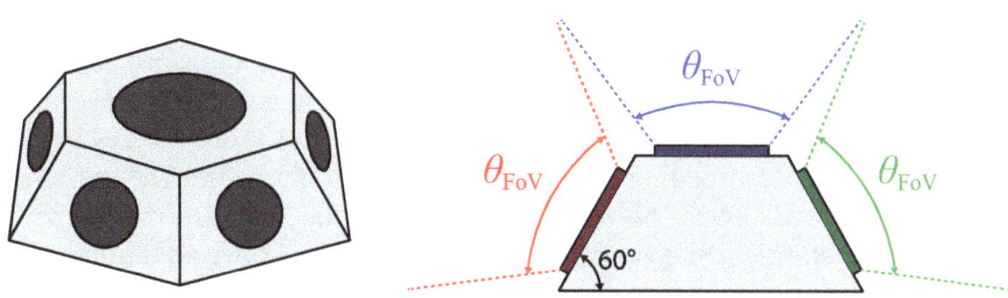

Figure 3.16. The multi-PD angle diversity receiver proposed in [57].

directionality of the incoming link due to an increase in SNR through selecting the strongest component [57]. This method also offers several other advantages including robustness against blocking and shadowing, where the link is fully or partially covered, respectively, as the receiver would automatically switch if another link suddenly offers a superior optical power.

In [49, 58], dynamic environments are proposed that consider the movement of human beings, alongside the usual concerns that shadowing and/or blocking can be an impediment to the link quality. It is shown in [49] through simulations that the received signal is rarely lost regardless of human movement, and the link was recoverable up to 98% of the time, while the optical power penalty due to shadowing was also insignificant and did not majorly impact the link in up to 99% of simulations [58].

When there are people present and mobile in the room, the VLC channel state information changes, and hence the channel bandwidth and SNR can vary accordingly. The same multi-angle approach that was used in [57] can also be used to mitigate this effect through the same means as previously explained. In [49], a similar but more thorough analysis was performed by simulation where the authors investigated several different scenarios including the movement of people across a room.

Based on the equations defined above, in particular (3.27), which defines the overall temporal channel impulse response including the bandwidth limitations introduced by multi-path effects, one must consider the movement of people through the system, since this will vary the parameter N_r and also change the mean excess delay and RMS delay spread, based on reflections from people in a dynamic manner, and hence, will impact the link performance.

To give an idea of the impact of natural human movement on the VLC link, the human body was modelled as a bulk cuboid with dimensions $1.8 \times 0.3 \times 0.2$ m. The bulk cuboid was allowed to move at a constant speed ranging from 2 to 5 km/h and each cuboid starts randomly distributed within the simulation environment. The authors defined three separate rooms for consideration. The first is a long, unfurnished corridor that has dimensions of $3 \times 20 \times 3$ (width $\times$ length $\times$ height). The second is a furnished office room with dimensions ($6 \times 7 \times 3$) and desk objects are spaced sporadically through the room, as illustrated in figure 3.17(a), while all the intricate dimensions are illustrated in figure 3.17(b). The final room investigated was an empty hall with dimensions ($8 \times 12 \times 3$).

The number of people selected in each case was kept the same at 0.16–0.17 people/m^2 and is set relative to the number of spaces that are available in the furnished office environment, which is limited to 7. Each of the simulated people (cuboids) can never exist in the same position and were set to move randomly for 30 s. As each of the simulated people moves around each of the rooms, the key performance indicators are recorded at discrete one second intervals. The parameters recovered are received optical power and the RMS delay spread. From these two, SNR, bandwidth and the ISI span can be derived from the above equations.

The room scenario considers both a LOS and multiple NLOS paths where appropriate. The simulated LEDs were grouped into clusters of 100×100 and had

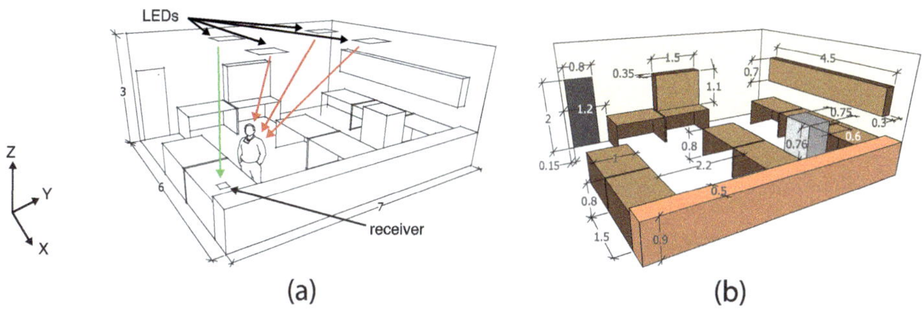

Figure 3.17. The room scenario considered in [49], where (a) shows the LED layout and overall room dimensions, whereas (b) shows the particular dimensions of the furniture within the room. © 2020 IEEE. Reprinted, with permission, from [49].

Table 3.3. The parameters considered for each scenario of [49].

Parameter	Location	Dimensions
Room size	Corridor	(20 × 3 × 3) m
	Office	(6 × 7 × 3) m
	Hall	(12 × 8 × 3) m
Position of LEDs	Corridor	(2.5, 1.5), (7.5, 1.5), (12.5, 1.5), (17.5, 1.5)
	Office	(2, 2), (4, 2), (2, 5), (4, 5)
	Hall	(4, 2), (8, 2), (4, 6), (8, 6)

10 mW power output. The clusters were evenly distributed in regular intervals across the room spacings. The room dimensions and spacings are confirmed in table 3.3 and were selected in the work to ensure consistency with the literature [49, 59, 60].

Each of the multi-path reflections were assumed to be Lambertian in nature, i.e. it is an absorption and re-emission with a Lambertian pattern, and only a single reflection is considered since the vast majority of power is lost beyond the second reflection. The reflection coefficient ρ was set to 0.8 [61]. In the order presented in the table, the paper started with the corridor. Scaling the people density of 0.16 people/m^2 according to the space, the paper assumed there would be 10 people in the corridor (0.16 people/m^2 × 20 m × 3 m ≈ 10 people). The intensity map describing the simulated movement of people is illustrated in figure 3.18, which shows, as expected, that the people mainly pass through the corridor and exit at either end.

The cumulative distribution functions (CDFs) of the received power and RMS delay spread are illustrated in figure 3.19 with and without people. There is a relatively minor difference in the two curves, and it could generally be suggested that the movement of people through the corridor generally doesn't heavily impact the performance of the VLC link. The reason for this is because there is always a LOS link between any of the LED clusters, which is the dominant path, in combination

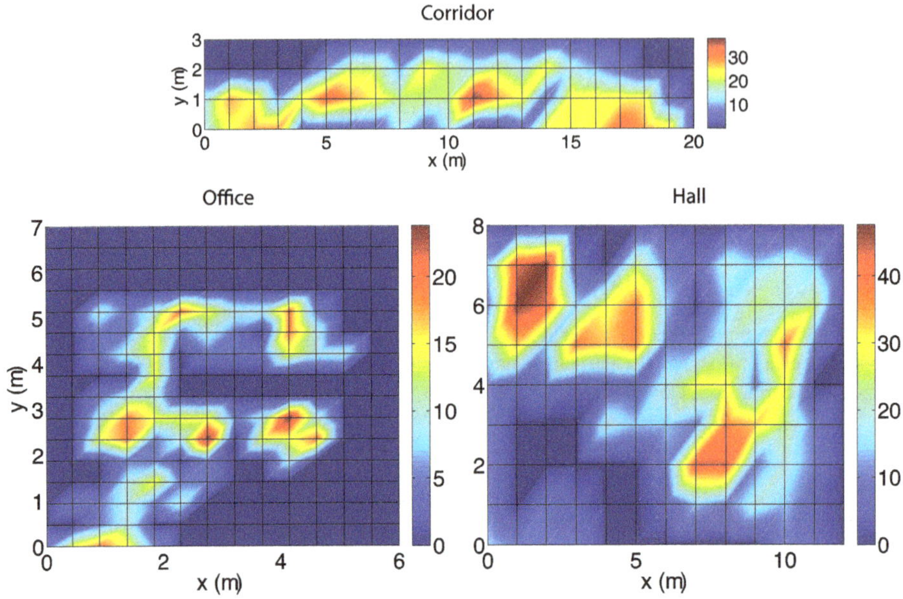

Figure 3.18. The simulated intensity map for people passing through each type of room tested, the hall, office and corridor. © 2020 IEEE. Reprinted, with permission, from [49].

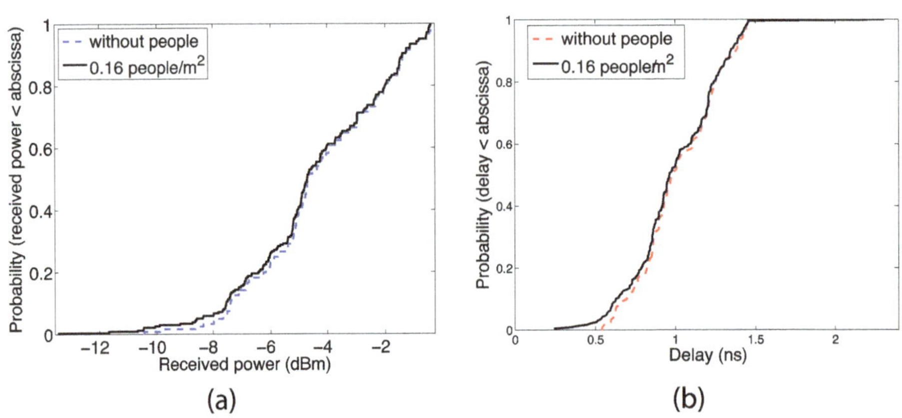

Figure 3.19. The CDFs of the (a) received power and (b) RMS delay spread for the corridor with and without people. © 2020 IEEE. Reprinted, with permission, from [49].

with the long, thin geometry of the room, meaning that there is very little difference introduced even if the people block the signal from one cluster.

For the furnished office, recalling that there were seven people modelled with the furniture and dimensions shown in figures 3.17(a) and (b), the CDFs are illustrated in figure 3.20 and immediately it is possible to notice that there is an obvious difference in comparison to the corridor. The difference is that the curves relating to the populated office have shifted to the left. This means that the received power has

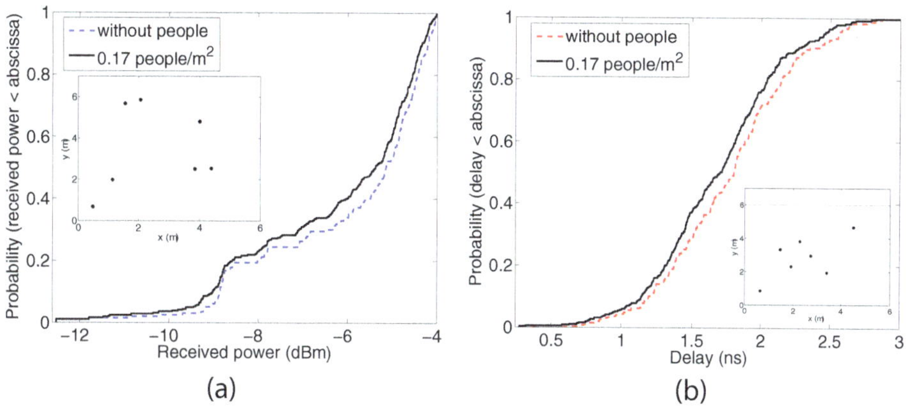

Figure 3.20. The CDFs of the (a) received power and (b) RMS delay spread for the office with and without people. © 2020 IEEE. Reprinted, with permission, from [49].

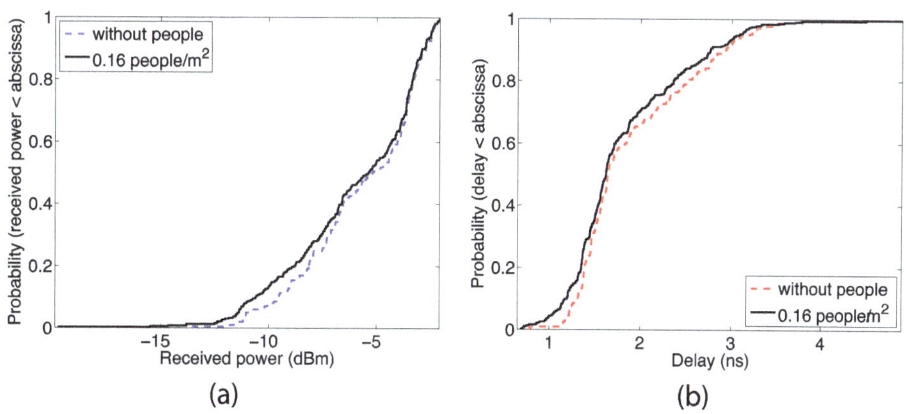

Figure 3.21. The CDFs of the (a) received power and (b) RMS delay spread for the hall with and without people. © 2020 IEEE. Reprinted, with permission, from [49].

dropped in comparison to the empty room and also the RMS delay spread is lower. The fact that the power drops is obvious, since from time-to-time, the cuboid simulation of humans will interrupt the light-path and therefore the light distribution occurs at slightly lower values. Less intuitive, however, is the delay spread dropping, which is advantageous, since a lower delay results in lower ISI span. The reason the delay drops is because the humans tend not to disturb the LOS between the transmitter and receiver and more often they do disturb the reflected paths. Having the reflected paths blocked means that the RMS average time-of-arrival of the signals is lower, since the long paths are more often blocked by the humans.

Finally, for the empty hall, 15 people are simulated ($0.16\ \text{people/m}^2 \times 12\ \text{m} \times 8\ \text{m} \approx 15$ people), maintaining the same distribution as previously, see figure 3.21.

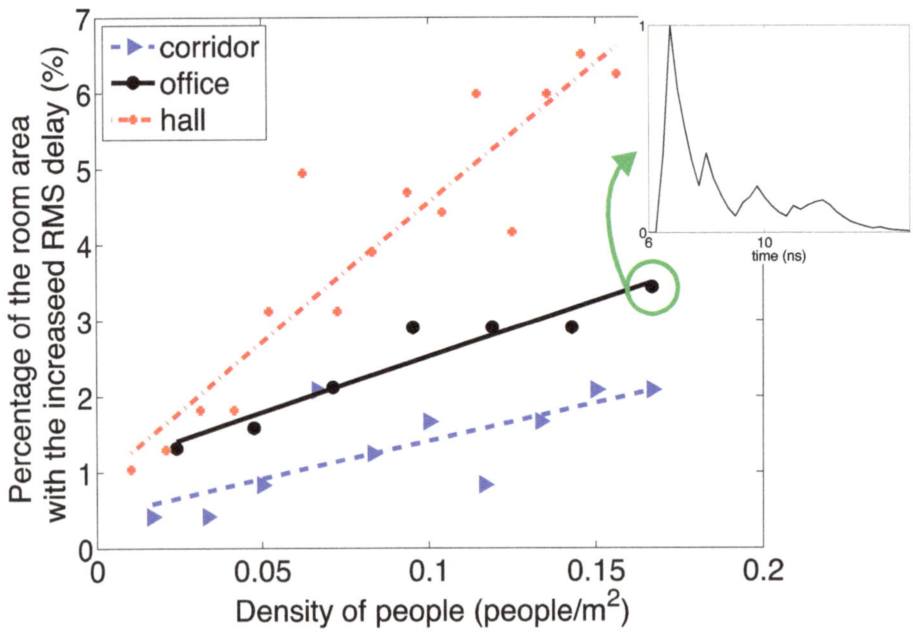

Figure 3.22. A comparison of how the density of people impacts the RMS delay spread. Clearly as more people enter the room there is a demonstrable increase in the delay. © 2020 IEEE. Reprinted, with permission, from [49].

The same overall effect as the furnished office is found, with the population curves shifting to the left. Clearly, when the number of paths is unrestricted, this is a general trend that occurs.

To understand the influence of the population density, the report shows the impact of having the rooms under test sparsely populated. While the CDF of the RMS delay spread has decreased overall for the specific population density above, it is not the whole picture for the entire area of the room. Figure 3.22 illustrates that there are areas of each of the rooms that have an increased RMS delay spread, particularly when the population density increases, in comparison to the unpopulated case. The corridor showed a small increase while the empty hall shows a relatively large increase. This is due to the fact that there are some areas where the LOS paths may be blocked and/or two or more multi-path components may have sufficient power, arriving with a significant delay, as can be seen in the inset, where the secondary peaks represent relatively strong multi-path components that increase the RMS delay spread.

3.4 Summary

This chapter has focused on high-speed circuit design and channel modelling. Several propositions of circuits reported in the literature were discussed starting with multiple-resonant equalisation, which equalises the low-pass response of an LEDs matrix by adding additional driving circuits to equalise a small portion of each LED in the matrix. The aggregated response of this results in an extended bandwidth by compensation of the low frequency components. However, the disadvantage of this method and any similar resistive-based equaliser is that the light output is reduced significantly causing a drop in SNR. Simultaneously, adding large blocking capacitors into the system also introduces a baseline wander effect, causing signal instability.

Therefore, an artificial transmission line synthesis approach was presented next, which aimed to cleverly distribute the LED capacitance into inductance added in parallel with the device. This method still has a resistive effect, as there must be a load resistor in series with the added inductance; however, a huge improvement of bandwidth could be obtained.

An even more useful method that is difficult to implement is negative impedance conversion, and the operating principles of that were outlined here through transfer equations and circuit diagrams. Negative impedance converters effectively project a negative capacitance in parallel with the LED, and as a result the net capacitance tends towards zero. The advantage of this system, which is significant, is that there is minimal resistive loading due to the net negative impedance and therefore there is no optical power penalty associated, in contrast to most other analogue equalisation schemes.

Next, channel models were considered, which are an important aspect that should be properly modelled. A number of different channel configurations were presented with a focus on the LOS configuration that is the most common. The radiation patterns of LEDs were discussed that have a significant impact on overall performance, as directed links tend to have higher SNR and bit rates, but lower mobility. Finally, a well-cited report in the literature that discusses channel models with people present in the office was analysed.

References

[1] Le Minh H, O'Brien D, Faulkner G, Zeng L, Lee K, Jung D and Oh Y 2008 80 Mbit/s visible light communications using pre-equalized white LED *34th European Conf. on Optical Communication, 2008. ECOC 2008* (IEEE) pp 1–2

[2] Le Minh H, O'Brien D, Faulkner G, Zeng L, Lee K, Jung D and Oh Y 2008 High-speed visible light communications using multiple-resonant equalization *IEEE Photonics Technol. Lett.* **20** 1243–5

[3] Tsiatmas A, Baggen C P M, Willems F M J, Linnartz J-P M G and Bergmans J W M 2014 An illumination perspective on visible light communications *IEEE Commun. Mag.* **52** 64–71

[4] Zhang S, Watson S, McKendry J J D, Massoubre D, Cogman A, Gu E, Henderson R K, Kelly A E and Dawson M D 2013 1.5 Gbit/s multi-channel visible light communications using CMOS-controlled GaN-based LEDs *J. Lightwave Technol.* **31** 1211–6

[5] Li H, Chen X, Huang B, Tang D and Chen H 2014 High bandwidth visible light communications based on a post-equalization circuit *IEEE Photonics Technol. Lett.* **26** 119–22

[6] Mirvakili A and Koomson V J 2012 High efficiency LED driver design for concurrent data transmission and PWM dimming control for indoor visible light communication *Photonics Society Summer Topical Meeting Series, 2012 IEEE* (IEEE) pp 132–3

[7] Zhao S, Xu J and Trescases O 2013 A dimmable LED driver for visible light communication (VLC) based on LLC resonant DC-DC converter operating in burst mode *2013 28th Annual IEEE Applied Power Electronics Conf. and Exposition (APEC)* pp 2144–50
[8] Huang X, Shi J, Li J, Wang Y, Wang Y and Chi N 2015 750Mbit/s visible light communications employing 64QAM-OFDM based on amplitude equalization circuit *Optical Fiber Communication Conf.* (Optical Society of America) p Tu2G-1
[9] Huang X, Shi J, Li J, Wang Y and Chi N 2015 A Gb/s VLC transmission using hardware preequalization circuit *IEEE Photonics Technol. Lett.* **27** 1915–8
[10] Haigh P A, Ghassemlooy Z, Rajbhandari S and Papakonstantinou I 2013 Visible light communications using organic light emitting diodes *IEEE Commun. Mag.* **51** 148–54
[11] Mirvakili A, Koomson V J, Rahaim M, Elgala H and Little T D C 2015 Wireless access test-bed through visible light and dimming compatible OFDM *2015 IEEE Wireless Communications and Networking Conf. (WCNC)* (IEEE) pp 2268–72
[12] Mirvakili A and Koomson V J 2014 A flicker-free CMOS LED driver control circuit for visible light communication enabling concurrent data transmission and dimming control *Analog Integr. Circ. Sig. Process.* **80** 283–92
[13] Chow C-W, Yeh C H, Liu Y F and Liu Y 2011 Improved modulation speed of LED visible light communication system integrated to main electricity network *Electron. Lett.* **47** 867–8
[14] Kumar K S S 2008 *Electric Circuits and Networks* (New Delhi: Pearson Education India)
[15] Street A M, Samaras K, O'Brien D C and Edwards D J 1997 Closed form expressions for baseline wander effects in wireless IR applications *Electron. Lett.* **33** 1060–2
[16] Odedeyi T, Haigh P A and Darwazeh I 2018 Transmission line synthesis approach to extending the bandwidth of LEDs for visible light communication *2018 11th Int. Symp. on Communication Systems, Networks Digital Signal Processing (CSNDSP)* pp 1–4
[17] Baeyens Y, Weimann N, Houtsma V, Weiner J, Yang Y, Frackoviak J, Roux P, Tate A and Chen Y K 2006 Submicron InP D-HBT single-stage distributed amplifier with 17 dB gain and over 110 GHz bandwidth *2006 IEEE MTT-S Int. Microwave Symp. Digest* (IEEE) pp 818–21
[18] Wong T T Y 1993 *Fundamentals of Distributed Amplification* (London: Artech House)
[19] Eriksson K, Darwazeh I and Zirath H 2015 InP DHBT distributed amplifiers with up to 235-GHz bandwidth *IEEE Trans. Microw. Theory Tech.* **63** 1334–41
[20] Odedeyi T, Giannakopoulos S, Zirath H and Darwazeh I 2019 Single-stage and multi-plicative distributed amplifiers for 200 GHz+ amplification *2019 IEEE Asia-Pacific Microwave Conf. (APMC)* pp 640–2
[21] Odedeyi T O 2020 *Distributed Circuit Analysis and Design for Ultra-Wideband Communication and Sub-mm Wave Applications* (London: University College London)
[22] Tanaka H, Umeda Y and Takyu O 2011 High-speed LED driver for visible light communications with drawing-out of remaining carrier *2011 IEEE Radio and Wireless Symp.* (IEEE) pp 295–8
[23] Liao C-L, Chang Y-F, Ho C-L and Wu M-C 2013 High-speed GaN-based blue light-emitting diodes with gallium-doped ZnO current spreading layer *IEEE Electron Device Lett.* **34** 611–3
[24] Poole C and Darwazeh I 2015 *Microwave Active Circuit Analysis and Design* (New York: Academic)

[25] Deng P, Kavehrad M and Kashani M A 2015 Nonlinear modulation characteristics of white LEDs in visible light communications *2015 Optical Fiber Communications Conf. and Exhibition (OFC)* pp 1–3
[26] Rasekh A and Bakhtiar M S 2016 Compensation method for multistage opamps with high capacitive load using negative capacitance *IEEE Trans. Circuits Syst. II: Express Briefs* **63** 919–23
[27] Han J, Yoo K, Lee D, Park K, Oh W and Park S M 2011 A low-power gigabit CMOS limiting amplifier using negative impedance compensation and its application *IEEE Trans. Very Large Scale Integr. (VLSI) Syst.* **20** 393–9
[28] Linvill J G 1954 RC active filters *Proc. IRE* 42 555–64
[29] Kassem A and Darwazeh I 2019 Exploiting negative impedance converters to extend the bandwidth of LEDs for visible light communication *2019 26th IEEE Int. Conf. on Electronics, Circuits and Systems (ICECS)* pp 302–5
[30] Filipkowski A 1999 Poles and zeros in transistor amplifiers introduced by Miller effect *IEEE Trans. Educ.* **42** 349–51
[31] Guerra O, Rodriguez-Garcia J D, Fernandez F V and Rodríguez-Vázquez A 2002 A symbolic pole/zero extraction methodology based on analysis of circuit time-constants *Analog Integr. Circ. Sig. Process.* **31** 101–18
[32] Huang X, Wang Z, Shi J, Wang Y and Chi N 2015 1.6 Gbit/s phosphorescent white LED based VLC transmission using a cascaded pre-equalization circuit and a differential outputs PIN receiver *Opt. Express* **23** 22034–42
[33] Fuada S, Putra A P, Aska Y and Adiono T 2017 A first approach to design mobility function and noise filter in VLC system utilizing low-cost analog circuits *Int. J. Recent Contribut. Eng. Sci. IT (iJES)* **5** 14–30
[34] Adiono T, Pradana A, Putra R V W and Fuada S 2016 Analog filters design in VLC analog front-end receiver for reducing indoor ambient light noise *2016 IEEE Asia Pacific Conf. on Circuits and Systems (APCCAS)* (IEEE) pp 581–4
[35] Fujimoto N and Mochizuki H 2013 477 Mbit/s visible light transmission based on OOK-NRZ modulation using a single commercially available visible LED and a practical LED driver with a pre-emphasis circuit *National Fiber Optic Engineers Conf.* (Optical Society of America) p JTh2A-73
[36] Li H, Chen X, Guo J, Tang D, Huang B and Chen H 2014 Mb/s visible optical wireless transmission based on NRZ-OOK modulation of phosphorescent white LED and a pre-emphasis circuit *Chin. Opt. Lett.* **12** 100604
[37] Burton A, Minotto A, Haigh P A, Ghassemlooy Z, Le Minh H, Cacialli F and Darwazeh I 2019 Optoelectronic modelling, circuit design and modulation for polymer-light emitting diodes for visible light communication systems *2019 26th Int. Conf. on Telecommunications (ICT)* (IEEE) pp 55–9
[38] Kassem A and Darwazeh I 2019 A high bandwidth modified regulated cascode TIA for high capacitance photodiodes in VLC *2019 IEEE Int. Symp. on Circuits and Systems (ISCAS)* (IEEE) pp 1–5
[39] Shannon C E 2001 A mathematical theory of communication *ACM SIGMOBILE Mob. Comput. Commun. Rev.* **5** 3–55
[40] Al-Kinani A, Wang C-X, Zhou L and Zhang W 2018 Optical wireless communication channel measurements and models *IEEE Commun. Surv. Tutor.* **20** 1939–62

[41] Barry J R 2012 *Wireless Infrared Communications* vol 280 (Berlin: Springer)
[42] Kahn J M and Barry J R 1997 Wireless infrared communications *Proc. IEEE* **85** 265–98
[43] Ghassemlooy Z, Popoola W and Rajbhandari S 2019 *Optical Wireless Communications: System and Channel Modelling with Matlab®* (Boca Raton, FL: CRC Press)
[44] Gupta A and Garg P 2018 Statistics of SNR for an indoor VLC system and its applications in system performance *IEEE Commun. Lett.* **22** 1898–901
[45] Stefan I, Elgala H and Haas H 2012 Study of dimming and LED nonlinearity for ACO-OFDM based VLC systems *2012 IEEE Wireless Communications and Networking Conf. (WCNC)* (IEEE) pp 990–4
[46] Barry J R, Kahn J M, Krause W J, Lee E A and Messerschmitt D G 1993 Simulation of multipath impulse response for indoor wireless optical channels *IEEE J. Sel. Areas Commun.* **11** 367–79
[47] Lee K, Park H and Barry J R 2011 Indoor channel characteristics for visible light communications *IEEE Commun. Lett.* **15** 217–9
[48] Gfeller F R and Bapst U 1979 Wireless in-house data communication via diffuse infrared radiation *Proc. IEEE* **67** 1474–86
[49] Chvojka P, Zvanovec S, Haigh P A and Ghassemlooy Z 2015 Channel characteristics of visible light communications within dynamic indoor environment *J. Lightwave Technol.* **33** 1719–25
[50] Hashemi H and Tholl D 1994 Statistical modeling and simulation of the RMS delay spread of indoor radio propagation channels *IEEE Trans. Vehicular Technol.* **43** 110–20
[51] Carruthers J B, Carroll S M and Kannan P 2003 Propagation modelling for indoor optical wireless communications using fast multi-receiver channel estimation *IEE Proc. Optoelectron.* **150** 473–81
[52] Holloway C L, Shah H A, Pirkl R J, Remley K A, Hill D A and Ladbury J 2012 Early time behavior in reverberation chambers and its effect on the relationships between coherence bandwidth, chamber decay time, RMS delay spread, and the chamber buildup time *IEEE Trans. Electromagn. Compat.* **54** 714–25
[53] Ghassemlooy Z, Alves L N, Zvanovec S and Khalighi M-A 2017 *Visible Light Communications: Theory and Applications* (Boca Raton, FL: CRC Press)
[54] Wu D, Ghassemlooy Z, Le Minh H, Rajbhandari S and Boucouvalas A C 2012 Improvement of the transmission bandwidth for indoor optical wireless communication systems using a diffused gaussian beam *IEEE Commun. lett.* **16** 1316–9
[55] Grubor J, Randel S, Langer K-D and Walewski J W 2008 Broadband information broadcasting using LED-based interior lighting *J. Lightwave Technol.* **26** 3883–92
[56] Ramirez-Aguilera A M, Luna-Rivera J M, Guerra V, Rabadán J, Perez-Jimenez R and Lopez-Hernandez F J 2018 A review of indoor channel modeling techniques for visible light communications *2018 IEEE 10th Latin-American Conf. on Communications (LATINCOM)* (IEEE) pp 1–6
[57] Burton A, Le Minh H, Ghassemlooy Z, Rajbhandari S and Haigh P 2012 A performance analysis for 180°; receiver in visible light communications *2012 4th Int. Conf. on Communications and Electronics (ICCE)* pp 48–53
[58] Jivkova S and Kavehrad M 2003 Shadowing and blockage in indoor optical wireless communications *Global Telecommunications Conf., 2003. GLOBECOM '03. IEEE* vol 6 pp 3269–73

[59] Komine T, Haruyama S and Nakagawa M 2005 A study of shadowing on indoor visible-light wireless communication utilizing plural white LED lightings *Wirel. Pers. Commun.* **34** 211–25

[60] Biagi M, Borogovac T and Little T D C 2013 Adaptive receiver for indoor visible light communications *J. Lightwave Technol.* **31** 3676–86

[61] Komine T and Nakagawa M 2004 Fundamental analysis for visible-light communication system using LED lights *IEEE Trans. Consum. Electron.* **50** 100–7

Chapter 4

Applications of signal processing

4.1 Introduction

One of the key technologies that supports VLC is digital signal processing (DSP), which is used across almost every single aspect of the end-to-end link. This chapter will cover the application of DSP in VLC starting with the advanced modulation formats that form the physical layer of data transmission. This text assumes that the reader has the fundamental knowledge of the basic concepts in advanced modulation formats such as the Nyquist sampling frequency, constellation mapping, basic filter theory and error vector magnitude (EVM). If the reader does not possess this fundamental information, the information is presented in full in the outstanding texts [1, 2].

Modulation formats are important because they control the number of bits/symbol that can be transmitted at any given time. The best modulation formats are designed for their application and should provide sufficient spectral efficiency whilst being as power efficient as possible. The maximum rate that can be achieved was defined by Shannon [3]. A number of common modulation formats are shown in figure 4.1.

4.2 Advanced modulation formats

Standard modulation formats such as pulse amplitude modulation (PAM), pulse position modulation (PPM) and general pulse-based schemes have been widely covered in the literature and will not be treated here [1]. However, there are a series of advanced modulation formats that are becoming increasingly popular in VLC systems such as OFDM or CAP. These formats will be discussed along with other emerging and popular formats such as multi-band carrier-less amplitude and phase modulation (*m*-CAP), staggered carrier-less amplitude and phase modulation (s-CAP) and fast orthogonal frequency division multiplexing FOFDM. The former is a system where numerous low-rate orthogonally spaced subcarriers are aggregated to form a high-rate system. By spacing the subcarriers orthogonally, they don't

doi:10.1088/978-0-7503-1680-4ch4

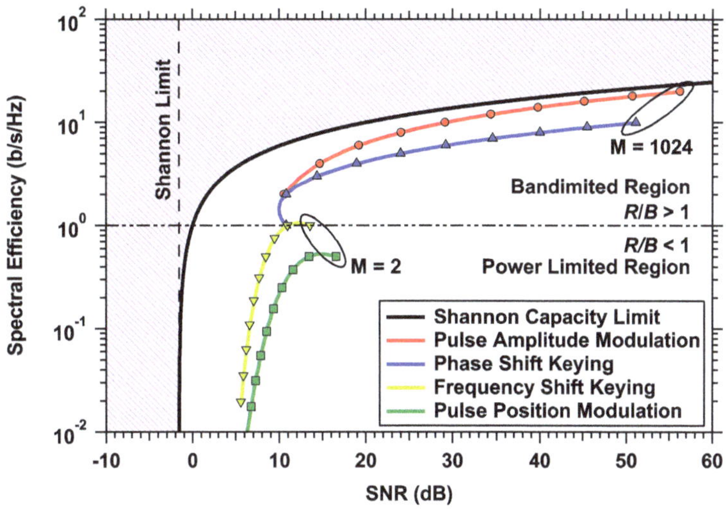

Figure 4.1. Spectral efficiency of a number of common modulation formats.

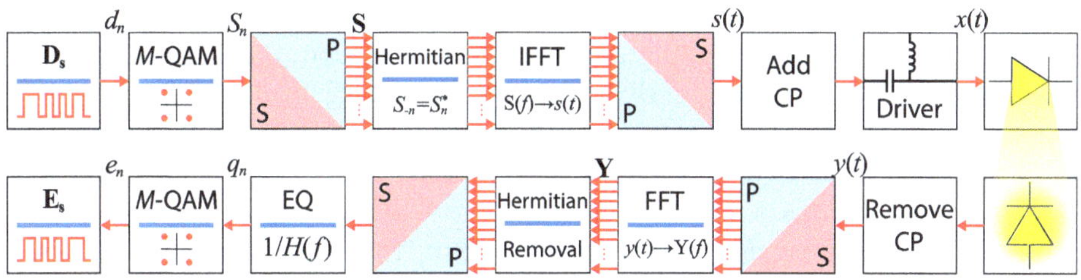

Figure 4.2. Block diagram for a Hermitian-symmetric OFDM system typically used in VLC.

interfere with each other and therefore each subcarrier can carry independent information [4, 5].

4.2.1 Orthogonal frequency division multiplexing

One of the most popular modulation formats used in VLC is OFDM, because it divides the available bandwidth into numerous low data rate subcarriers. This means that each subcarrier has its own small sub-bandwidth that can be used to carry information whilst offering protection from detrimental channel effects such as fading and attenuation. In general, OFDM uses an inverse fast Fourier transform (IFFT) to convert frequency-domain symbols into the time domain at the transmitter, and a fast Fourier transform (FFT) to recover the signal at the receiver.

A simplified, generic OFDM block diagram is illustrated in figure 4.2. A pseudo-random binary sequence (PRBS) sequence $\mathbf{D}_s = [d_0, d_1, \cdots, d_{N-1}]$ is generated and mapped onto the quadrature amplitude modulation (QAM) constellation, where the nth symbol is given by S_n, arriving at a rate of R_s. They are then converted into a vector $\mathbf{S} = [0, S_1, \cdots, S_{N/2}]$ consisting of $N/2$ parallel streams, each with a subcarrier speed $r_s = R_s/N$. The first index is a 0 because it is the subcarrier placed on the dc frequency. Importantly, only $N/2$ symbols are generated because of the requirement to satisfy the Hermitian symmetry condition to generate real numbers. This means that half of the subcarriers must be assigned the complex conjugate of their negative frequency counterpart, i.e. $S_{-n} = S_n^*$. This ensures that the IFFT outputs real numbers that are compatible with LEDs, which operates on the principle of IM. The continuous time IFFT is given by [5]:

$$s(t) = \frac{1}{\sqrt{T_s}} \sum_{n=0}^{N-1} S_n \exp\left[\frac{j2\pi nt}{T_s}\right] \tag{4.1}$$

for the nth subcarrier at instantaneous time instant t, $T_s = 1/R_s$ is the OFDM symbol period. Since the vast majority of modern communications systems are digital, it is necessary to convert the continuous time formulation of the IFFT into the discrete time version as follows [6]:

$$s(k) = \frac{1}{\sqrt{\gamma N}} \sum_{n=0}^{N-1} S_n \exp\left[\frac{j2\pi nk}{\gamma N}\right] \tag{4.2}$$

for the kth sample instance, and an oversampling factor of γ is introduced. Each of the N inputs to the IFFT modulate an individual frequency bin separated by an integer multiple of $1/T_s$ to maintain orthogonality. In figure 4.3(a), a system with $N = 5$ subcarriers is illustrated, as well as the overall OFDM envelope (dashed black line). The envelope of an OFDM system becomes a more ideal rectangular shape with increasing N. This is highlighted in figure 4.3(b) for $N = \{5, \ 64 \text{ and } 2048\}$ subcarriers. After conversion to the time domain, the samples must be serialised before a cyclic prefix (CP) is added.

The CP is typically included as a mechanism to protect the useful information from multi-path effects [7, 8]. The CP is normally added to the start of the symbol and is copied from a fraction of the end of the symbol, as illustrated in figure 4.4. If

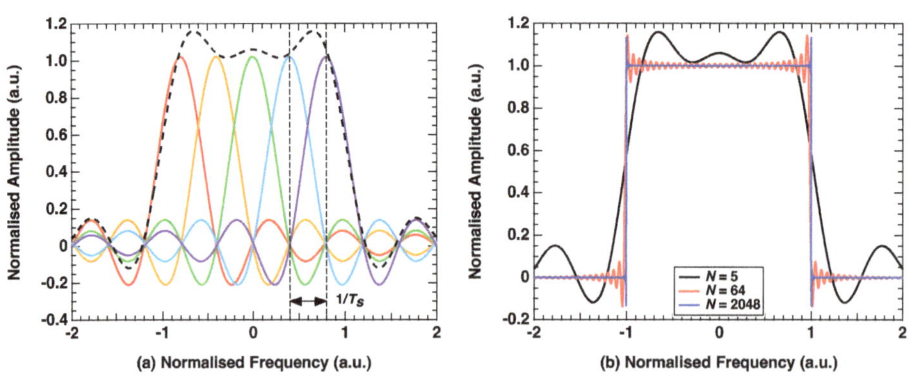

Figure 4.3. Increasing the number of subcarriers improves the spectral usage of an OFDM system but also increases its computational complexity. The systems shown have (a) 5 and (b) up to 2048 subcarriers. The signal envelopes clearly show that increasing the number of subcarriers improves the spectral usage by making it more square with a sharper cut-off.

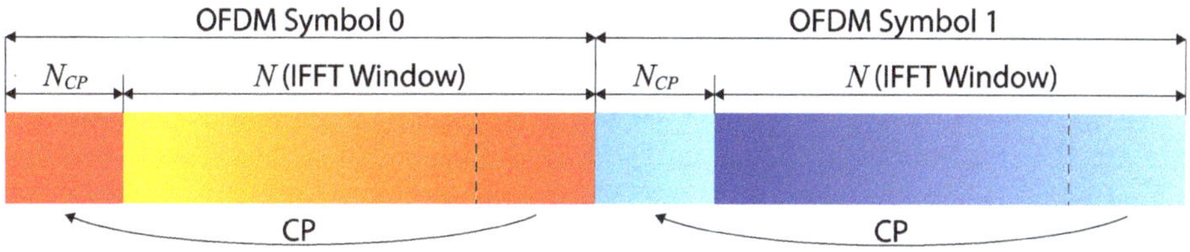

Figure 4.4. The concept of cyclic prefix; a portion of the end of the OFDM symbol is copied to the start of the same symbol to protect against multi-path fading and inter-symbol interference.

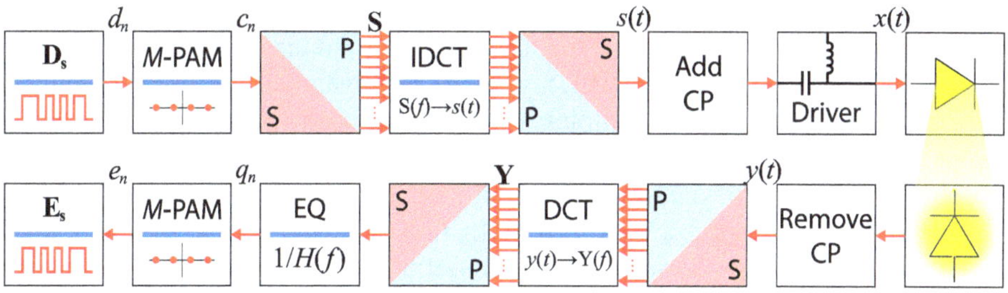

Figure 4.5. The block diagram for an FOFDM signal. The difference between this one and the one for conventional OFDM is the use of PAM (i.e. a one-dimensional modulation) and the IDCT instead of the IFFT. This means that Hermitian symmetry can be avoided with no loss of spectral efficiency.

the impulse response of the channel is sufficiently slow that it causes energy from one OFDM symbol to leak beyond the end of the symbol period, the CP is selected with adequate length to ensure that it does not interfere with the next.

After the CP is added, the signal is biased (driver technologies discussed in chapter 3) before intensity modulation of the LED. Transmission occurs over the optical channel, the models of which are also discussed in chapter 3 and a portion of the transmitted optical power is collected by the photodiode.

The first stage of OFDM demodulation is to remove the CP by simply deleting it. Next, the signal must be reverted into N-parallel streams in order to undergo the FFT process to convert the symbols back into the frequency domain, before removal of the Hermitian symmetry. The remaining symbols are then serialised once more before one-tap equalisation (EQ in figure 4.2), QAM constellation de-mapping and estimation of the transmitted signal vector $\mathbf{E}_s$. The concept of equalisers will be covered later in this chapter.

4.2.2 Fast orthogonal frequency division multiplexing

The so-called FOFDM is a special subset of OFDM, first introduced in 2002 by [9], the block diagram of which is illustrated in figure 4.5 and the key differences between FOFDM and OFDM are clearly the lack of Hermitian symmetry in the former, and both the modulation format used and the transform applied to generate the time-domain symbols.

In OFDM the subcarrier spacings are set at intervals of $1/T_s$ to maintain the orthogonality condition, as mentioned in the previous section. On the other hand, in

FOFDM the subcarrier spacings can be reduced to $1/2T_s$ in an effort to either double the number of subcarriers present in the system, or save half the bandwidth required for a given data rate. This is illustrated in figure 4.6. Instead of using the IFFT to modulate the symbols onto the subcarriers, an IDCT is used instead as follows [10, 11]:

$$s(k) = \frac{1}{\sqrt{\gamma N}} \sum_{n=0}^{N-1} S_n \cos\left\{\frac{j\pi nk}{\gamma N}\right\} \tag{4.3}$$

The trade-off for this bandwidth saving approach is that 1-dimensional real-valued modulation alphabets must be used, i.e. PAM and the quadrature component is lost due to IDCT that is used. In VLC systems, however, this is not a significant drawback because real-valued samples must be produced for the IM of the LEDs and thus, Hermitian symmetry can be avoided. Therefore, if an example system is set where an OFDM system has $N = 16$ subcarriers each modulated by 4-QAM (i.e. $k = 2$ bits/symbol) and a bandwidth $B = 1$ Hz, half of them must be used to load the conjugate data to satisfy the Hermitian symmetry condition. This means only half the bandwidth is used in a meaningful way and the total useful symbol rate is given by $R_s = Nr_s/2 = NkB_{sc}/2$ where B_{sc} is the subcarrier bandwidth given by $B_{sc} = B/N$. With the given conditions, $R_s = 1$ sym/s. While for FOFDM under the same conditions, i.e. 2-PAM, $N = 16$ to maintain the equivalent alphabet in a single dimension and subcarriers; however, since N is maintained, B_{FOFDM} becomes $B_{\text{FOFDM}} = B/2 = 0.5$. Since there is now no requirement for Hermitian symmetry, the overall symbol rate is given by $R_s = Nr_s = NkB_{sc}$ as before, but recalling that B_{sc} has reduced by half. Therefore, $R_s = 1$ sym/s and an equivalent data rate is maintained, even though half the bandwidth is used, as illustrated in figure 4.7.

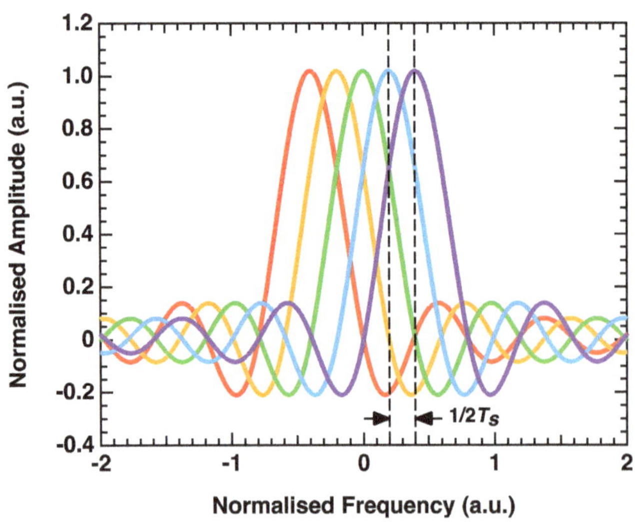

Figure 4.6. The FOFDM subcarrier assignment, showing that the cap between symbols has been reduced by half in comparison to OFDM.

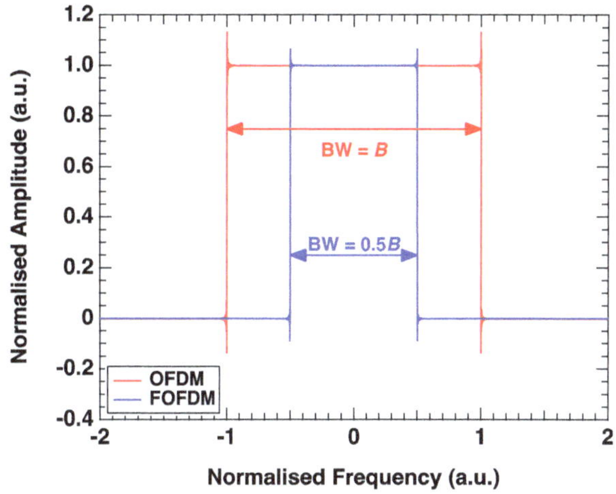

Figure 4.7. The overall comparative OFDM and FOFDM signal envelopes in the frequency domain. The two systems have equivalent spectral efficiency but FOFDM uses half the bandwidth. The reason for this is because it uses the IDCT instead of the IFFT, which means the subcarriers are spaced closer together and Hermitian symmetry can be avoided; however, the system is limited to one-dimensional modulation formats.

If the order of N is doubled, or if k is equivalent, the data rate will be doubled with no increase in system complexity. If both conditions are upheld, the data rate will be quadrupled with an increase in computational complexity by a factor of 2 due to the extra subcarriers. The former is particularly advantageous in VLC systems because the LEDs are band-limited. This means that maintaining an equivalent data rate but using half the bandwidth can yield some advantages, as was shown in the literature in [12], which shows that FOFDM offers a 3 dB reduction in SNR requirement in heavy band-limited environments.

4.2.3 Carrier-less amplitude and phase modulation

One of the main competitors to OFDM in VLC is CAP, which is similar in such that it is also a passband modulation and can also be thought of as similar to QAM in that it has a single spectral band, but does not use a local oscillator.

The block diagram for a generic CAP system is shown in figure 4.8. First, random data d of length s is buffered in the vector $\mathbf{D}_s = [d_0, d_1, \cdots, d_{s-1}]$ and then mapped into the desired M-QAM constellation before being split into its in-phase and quadrature components. Before it is possible to pulse shape the data, one must upsample the data to avoid aliasing in the filters. The rate of upsampling must be equal to the number of samples-per-symbol N_{samp}, which is given by [13]:

$$N_{\text{samp}} > 2(1 + \beta) \tag{4.4}$$

The origin of this is that the total symbol rate is fixed by the user. The pulse-shaping filters are typically generated with a root-raised cosine (RRC) as the basis function, which has an associated roll-off factor β. Hence the signal bandwidth after pulse shaping depends on β and the Nyquist sampling rate must also be satisfied. The

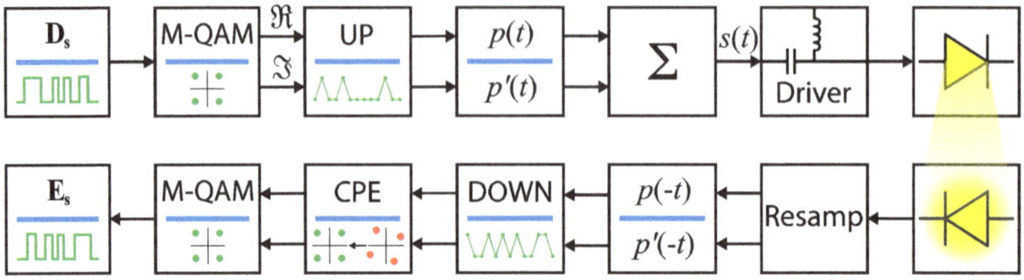

Figure 4.8. The CAP system block diagram. The overall block diagram is similar to that seen in QAM; however, the carrier frequency is inserted using pulse-shaping filters instead of local oscillators.

relationship between the symbol rate R_s and the signal bandwidth is therefore given by [13]:

$$B_{sig} = R_s(1 + \beta) \tag{4.5}$$

Therefore, to satisfy the Nyquist sampling rate conditions [14], the sampling frequency f_s must be set to the following [13]:

$$f_s > 2R_s(1 + \beta) \tag{4.6}$$

which, through substitution using (4.4) becomes [13]:

$$f_s = \gamma R_s N_{\text{samp}} \tag{4.7}$$

which is the most compact form. Here, a new term γ is introduced which represents an oversampling factor and has the condition $\gamma \in \mathbb{Z}^+$. If $\gamma = 1$ then the system is configured at the Nyquist rate.

After upsampling via zero-padding at a rate of N_{samp}, the in-phase and quadrature symbols must be shaped by their respective pulse-shaping filters given by $p(t)$ (in-phase) and $\bar{p}(t)$ (quadrature). The pulses are defined as the product between a basis function and a cosine and sine wave with carrier frequency f_c following [15, 16]:

$$p(t) = g(t)\cos\left(2\pi f_c t\right) \tag{4.8}$$

$$\bar{p}(t) = g(t)\sin\left(2\pi f_c t\right) \tag{4.9}$$

for the in-phase and quadrature, respectively, and $g(t)$ is a basis function that has historically been selected as a RRC, given by [17]:

$$g(t)\frac{\sin\left(\frac{\pi t}{T_s}[1-\beta]\right) + 4\beta\frac{t}{T_s}\cos\left(\frac{\pi t}{T_s}[1+\beta]\right)}{\frac{\pi t}{T_s}\left[1-\left(4\beta\frac{t}{T_s}\right)^2\right]} \tag{4.10}$$

The final transmitted signal $s(t)$ is given by [13]:

$$s(t) = \sqrt{2}[\Re\{a^i(t)\}\circledast p(t) + \Im\{a^i(t)\}\circledast \tilde{p}(t)] \tag{4.11}$$

where a^i is the ith complex QAM symbol and $\circledast$ represents time-domain convolution.

Then, as in the case of OFDM and every other modulation format, the signal is impressed onto the intensity of an LED via a driving circuit before photodetection and digitisation. The CAP demodulator is simply the opposite of the modulator. First, if any alteration has been made to the sampling frequency (i.e. via an oscilloscope) then resampling must occur before matched filtering using the time-reversed filters $p(-t)$ and $\tilde{p}(-t)$. Next, the signals are downsampled at the mid-point at a rate of N_{samp} before removal of common phase error (CPE) via phase rotation [18]. Then finally, constellation de-mapping is performed via a detection method which, in general, is hard thresholding before bit estimation $\mathbf{E}_s$.

4.2.4 Multi-band carrier-less amplitude and phase modulation

The conventional CAP format described above is not a multi-carrier system and hence is severely prone to attenuation introduced at high frequencies by the low modulation bandwidth of the LEDs. Hence, in [18], the author adopted the approach outlined in [13] to extend CAP into a multi-carrier system. The original extension to m-CAP where m is the number of sub-bands, was proposed by Monroy *et al*, and was intended for use in short-haul optical fibre links to combat chromatic dispersion. Clearly chromatic dispersion is not a serious problem in VLC and hence the application has a motivation that is embedded into the attenuation experienced because of the LEDs at high frequencies. It was postulated that by splitting the CAP signal into multiple sub-bands, the attenuation experienced by each sub-band would be lower, and hence, improved SNR performance could be obtained, resulting in a higher overall data rate, as has been reported in the literature for many different arrangements of the sub-bands [18–21].

The overall concept of m-CAP is illustrated in figure 4.9. The black line corresponds to a conventional CAP system outlined previously. The red, green and blue lines correspond to m-CAP systems with 2, 4 and 8 sub-bands, respectively. The left part of figure 4.9 shows the raw transmitted spectrum while the right side shows the effect of a low pass filter with normalised cut-off frequency of 0.5. Clearly as m increases in order, i.e. adding additional sub-bands to the system, the attenuation experienced by each sub-band is reduced as mentioned. One further advantage of reformulating CAP as a multi-band scheme is that each sub-band has its own SNR, meaning that each sub-band can carry a given number of bits that may be loaded adaptively according to the classical power and bit-loading principles [22].

The block diagram for a generic m-CAP system can easily be found by extending the CAP system previously outlined by additional sub-bands with increased carrier frequencies. This is highlighted in figure 4.10, where four pulse shapes are shown for a two sub-band system, starting with figure 4.10(a), which shows the first sub-band in-phase pulse shape (in red), which is found by the multiplication of the RRC

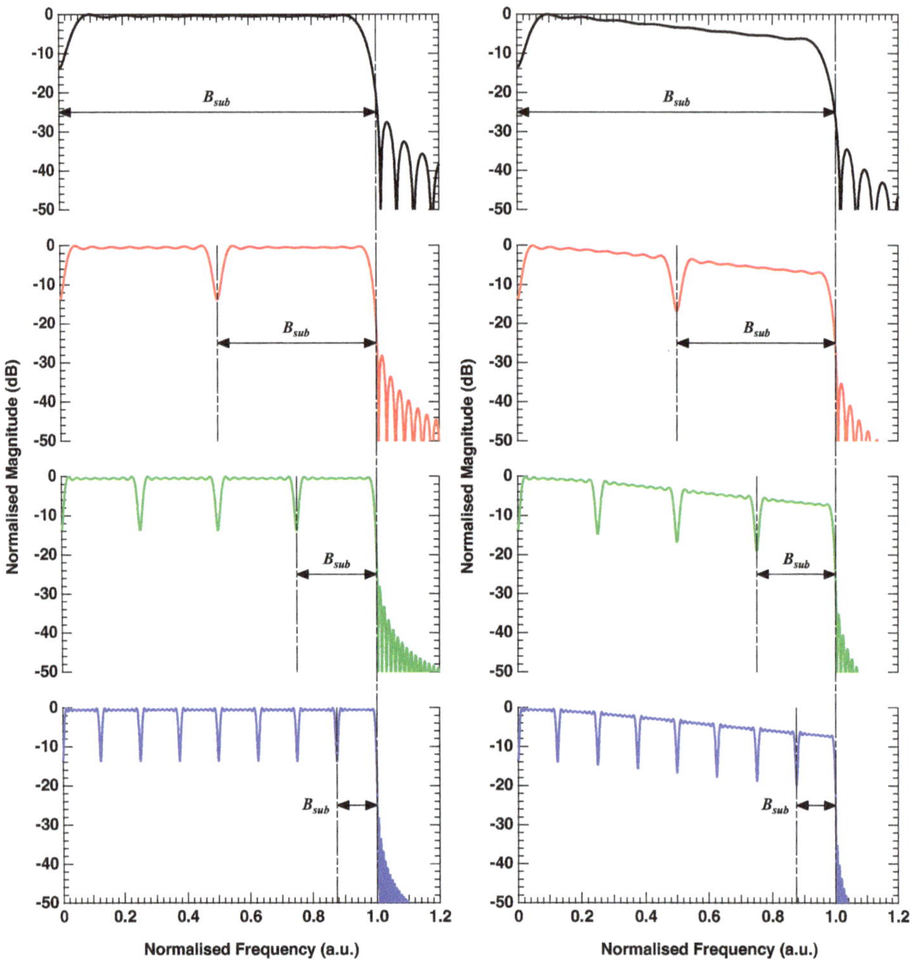

Figure 4.9. The m-CAP concept is shown for an increasing number of subcarriers on the left-hand side of the figure. On the right-hand side, it is shown that protection against high-frequency attenuation can be obtained by dividing the system into multiple sub-bands, at the cost of vastly increased spectral efficiency.

(black) with the cosine (grey) at carrier frequency f_{c1}. The quadrature pulse shape (blue) for the first sub-band is shown in figure 4.10(b), which is the product of the RRC and the sinusoid at f_{c1} (grey). Clearly, by inspection of figures 4.10(a)–(d), the only difference between the first and second sub-bands is the carrier frequency. In figures 4.10(c) and (d) the pulse shapes for the second sub-band are generated in exactly the same way as previously, albeit that the RRCs are multiplied by a cosine and sine wave carried at f_{c2} and obviously $f_{c2} > f_{c1}$.

The value of the nth sub-band in a generic m-band system can be found using the following relation [15]:

$$f_{cn} = \frac{B_{sig}(2n - 1)}{2m} \tag{4.12}$$

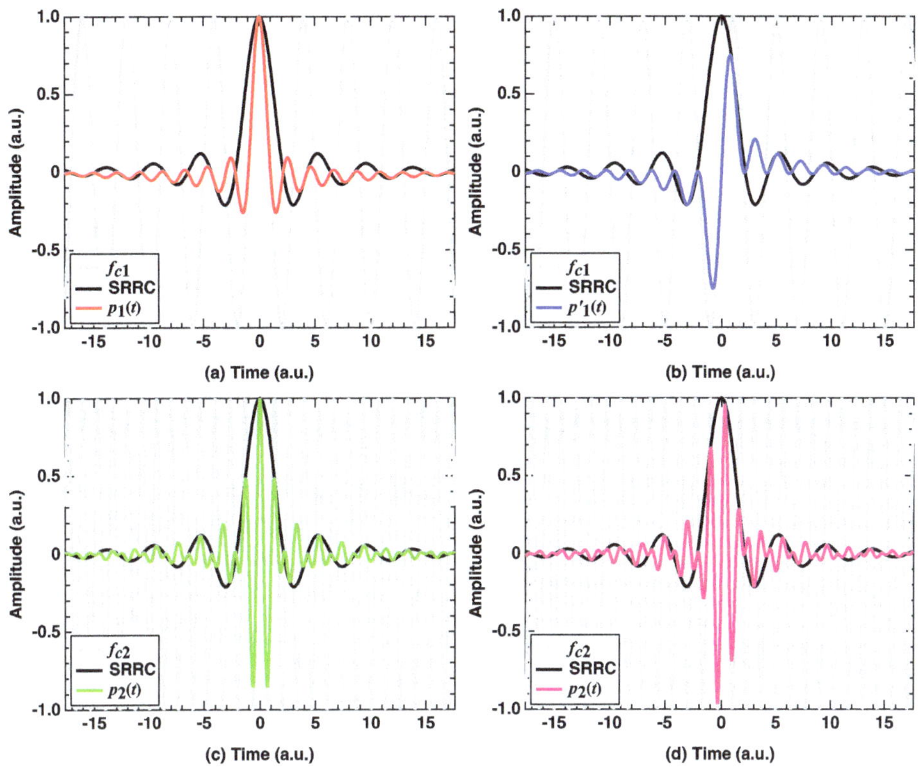

Figure 4.10. The pulse shapes used in the filters for the (a) in-phase and (b) quadrature components of the first sub-band, and (c) in-phase and (b) quadrature components of an arbitrary dual sub-band m-CAP system.

also recalling that the signal bandwidth is the total signal bandwidth. The pulse-shaping filters are given by the relationships outlined in the previous section on conventional CAP, however, one must be generated for each sub-band before pulse shaping.

The sampling rate is modified to the following [13]:

$$f_s = \frac{R_s N_{\text{samp}}}{m} \tag{4.13}$$

while N_{samp} becomes [13]:

$$N_{\text{samp}} = \gamma \lceil 2m(1 + \beta) \rceil \tag{4.14}$$

and used for upsampling via zero-padding as in single-carrier CAP. Next, the data corresponding with each sub-band is passed through the pulse-shaping filters (4.11) such that [13]:

$$s(t) = \sqrt{2} \sum_{n=1}^{m} \left[\Re\{a_n^i(t)\} \circledast p_n(t) + \Im\{a_n^i(t)\} \circledast \bar{p}_n(t) \right] \tag{4.15}$$

The demodulation of m-CAP is identical to that of CAP, however, it is simply repeated for every additional sub-band.

4.2.5 Staggered carrier-less amplitude and phase modulation

The problem with m-CAP modulation in general is the use of a roll-off factor that behaves like a guard-band in the frequency domain. Recently a new approach to CAP was proposed by Grzegorz Stepniak in [23] that offers complete spectral usage by inclusion of two additional PAM filters. This approach was denoted s-CAP and orthogonality between filters is maintained by introducing a $T/2$ offset between the original two CAP filters described in the previous sections and the two new filters following the concept of offset quadrature amplitude modulation (OQAM) [24, 25]. In [23, 26], BER tests were performed that show s-CAP outperforms 4-CAP at the same data rate.

The s-CAP system block diagram is illustrated in figure 4.11 and the main difference from CAP is that it is designed using four orthogonal filters; $f_0(n)$, $f_1(n)$, $f_2(n)$ and $f_3(n)$, where n is the current sample instance. The filters $f_1(n)$ and $f_2(n)$ are the Hilbert pair found in conventional CAP, i.e. they are given by (4.8) and (4.9). On the other hand, the additional PAM filter $f_0(n)$ is simply a RRC given by (4.10) and $f_3(n)$ is a RRC multiplied by a carrier as [23, 26]:

$$f_3(n) = g(n)\cos\left(2\pi f_{c3} n\right) \tag{4.16}$$

which are placed either side of $f_1(n)$ and $f_2(n)$, as illustrated in the frequency domain in figure 4.12 to ensure full spectral usage. The $f_3(n)$ carrier frequency f_{c3} is twice that of the carrier frequencies of the centre filters $f_1(n)$ and $f_2(n)$.

To maintain orthogonality between all filters, a $T/2$ offset is introduced between the $f_1(n)$ and $f_2(n)$. The reason for this is best explained with analysis of the filter cross-correlations, illustrated in figure 4.13. In figure 4.13(a), the cross-correlations of $f_1(n)$ with each of the other filters are shown. Clearly, there are several non-zero results at the interval $t = 0$ which will not allow successful transmission of data if left untreated. However, every cross-correlation has a zero-crossing when a $T/2$ symbol delay is introduced, as illustrated by the dashed lines. On the other hand, the remaining filter correlations are shown in figure 4.13 and they clearly all have a zero-crossing at the $t = 0$ interval. Therefore, introducing a $T/2$ shift in the data encoded

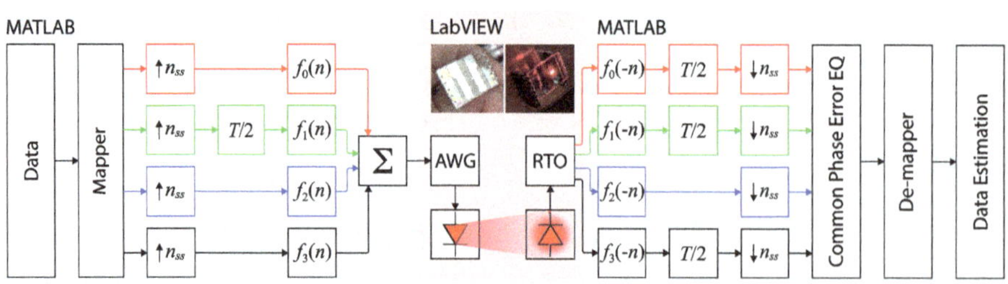

Figure 4.11. The staggered CAP system block diagram. © 2020 IEEE. Reprinted, with permission, from [26].

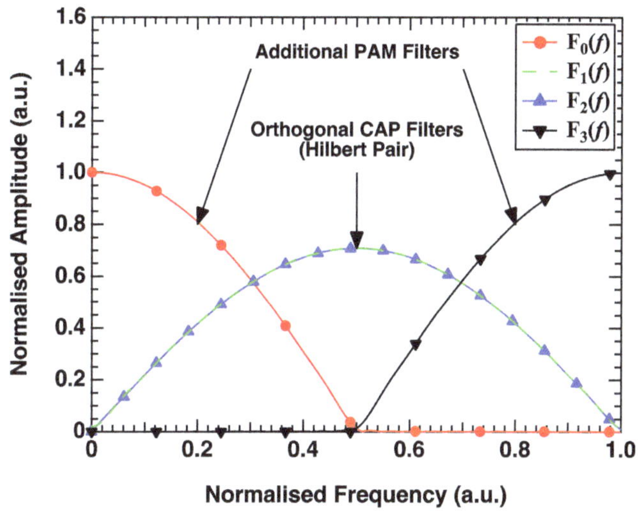

Figure 4.12. The frequency response of the staggered CAP pulse-shaping filters. © 2020 IEEE. Reprinted, with permission, from [26].

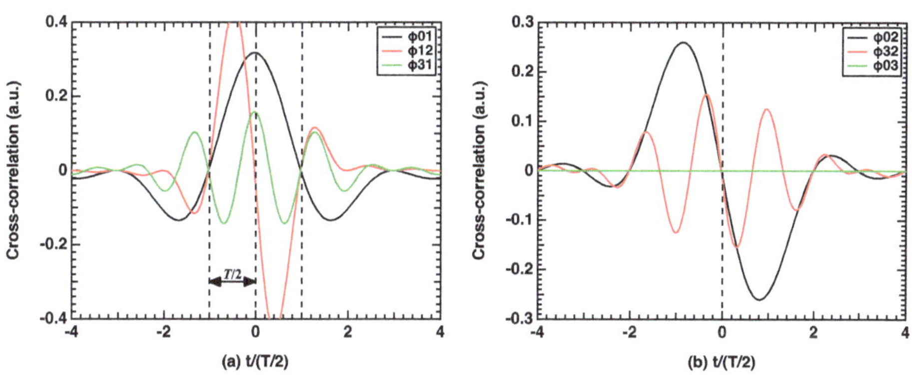

Figure 4.13. The cross-correlations of the relevant filters to show zero interference and therefore the recoverability of the signals.

by $f_1(n)$ will result in successful orthogonal co-existence and enable recovery of transmitted data.

The demodulation of s-CAP is slightly different to that of the other flavours of CAP described so far. The received signal is applied to the time-reversed matched filters $f_0(-n)$, $f_1(-n)$, $f_2(-n)$ and $f_3(-n)$ in the same way as previously, however, due to the timing offset introduced to maintain orthogonality, the $T/2$ symbol delay is used for three of the channels to ensure that the sampling point is consistent before downsampling. Following downsampling, CPE removal may be performed and symbol de-mapping can be performed to recover estimates of the original transmitted bits.

The BER performance of an s-CAP system in an experimental VLC link based on red PLEDs was reported in [26] in comparison to conventional 4-CAP and OOK. These results are illustrated in figure 4.14(a). Clearly, each of the sub-bands can support error free performance, provided each sub-band has a data rate beneath 900 kb/s. The reason for this is related to the modulation bandwidth of the PLED used, which was approximately 500 kHz. Further details can be found in [26], including the electro-optic response of the diode. When considering the 7% forward error correction (FEC) limit (i.e. a BER of 3.8×10^{-3}), the achievable sub-band rates range from approximately 1.7–1.4 Mb/s, depending on band placement. The first sub-band can support the highest data rate, because it is exposed to the least out-of-band attenuation as it occupies the lowest frequency range. The next two sub-bands are located in the same frequency space but are separated by π^c phase as in conventional CAP. Finally, the worst rates are obtained by the final channel, which is reliant on the matched filter, and hence the out-of-band attenuation distorts the signal shape, thus limiting the ability of the matched filter to remove the ISI.

A 4-CAP system with equivalent energy was also comparatively tested to compare its performance with s-CAP. The individual BER performances are illustrated in figure 4.14(b). Clearly, the BER performance of each sub-band is in the inverse order of sub-band frequency as the previous s-CAP results, for the same reasons. This means that the lowest frequency sub-band (i.e. $s = 1$) offers the best performance because of reduced impact of out-of-band attenuation. The achievable data rate is approximately 4.8 Mbs after summation of the sub-bands, which represents a reduction of over 1 Mb/s in comparison to s-CAP.

The total BER (i.e. the ratio of the total number of errors in all sub-bands to the total number of bits transmitted) for s-CAP, 4-CAP and OOK is illustrated in figure 4.15. The s-CAP scheme offers the un-coded rate at 3.75 Mb/s and 6.2 Mb/s when 7% FEC coding is considered. On the other hand, 4.6 Mb/s and 3.75 Mb/s are available for 4-CAP and OOK, respectively.

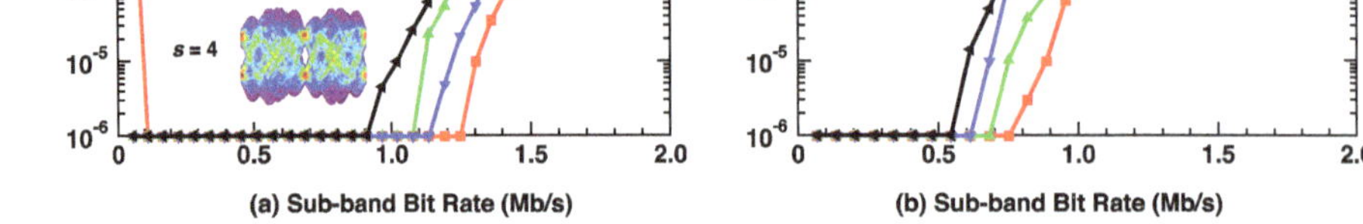

Figure 4.14. The experimental BER performance of each sub-band of the (a) staggered CAP system in comparison to the performance of a (b) conventional 4-CAP system. © 2020 IEEE. Reprinted, with permission, from [26].

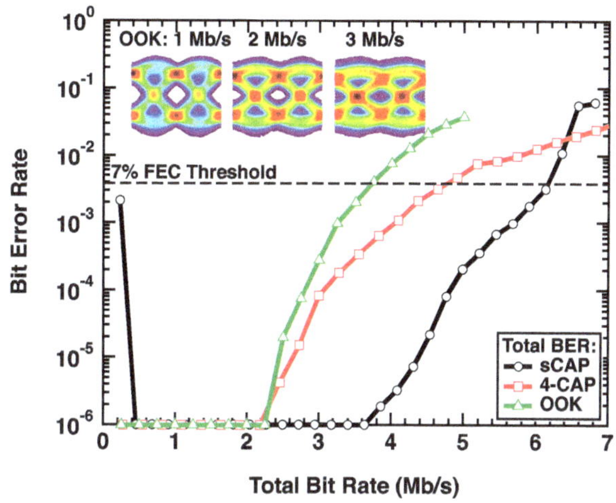

Figure 4.15. The overall experimental BER performance of (a) staggered CAP system in comparison to the performances of (b) conventional 4-CAP and (c) OOK. © 2020 IEEE. Reprinted, with permission, from [26].

4.2.6 Super-Nyquist carrier-less amplitude and phase modulation

Due to the popularity of *m*-CAP in VLC systems [18–20, 27, 28], alongside the simultaneous popularity of non-orthogonal modulation formats to increase spectral efficiencies, the super-Nyquist carrier-less amplitude and phase modulation (SNCAP) concept was generated simultaneously by Chi *et al* at Fudan University [29] and Haigh *et al* at University College London [30, 31] in 2018. The principle of SNCAP is to reduce the sub-band spacing, taking inspiration from SEFDM by intentionally violating the frequency-domain orthogonality between sub-bands to purposely introduce inter-band interference (IBI). For all intents and purposes, super-Nyquist is intended to mean the baud is far in excess of the bandwidth occupied in this paper, as in [32].

In SEFDM, as above, subcarrier spacing is compressed beyond the orthogonality limit (i.e. $<1/T$), resulting in increased spectral efficiency at the cost of deterministic inter-carrier interference (ICI) and increased computational complexity at the receiver. The time-domain equivalent is known as FTN signalling [33], where spectral efficiency is improved by intentional aliasing of pulse-shaping filters at the transmitter, introducing deterministic ISI. Since the interference generated in both cases is deterministic, knowledge of the correlation matrix enables a degree of recovery between the received and expected symbol values. Systems have been demonstrated that can increase throughput by up to 60% with a tolerable loss of performance [34–37].

The block diagram for SNCAP does not widely vary from that of conventional CAP as can be seen from figure 4.16. The only difference is in the sub-band carrier frequencies, where in (4.17) we introduce the term α that corresponds to the bandwidth saving by compression of the sub-bands following [37]. The carrier frequency for the nth sub-band becomes:

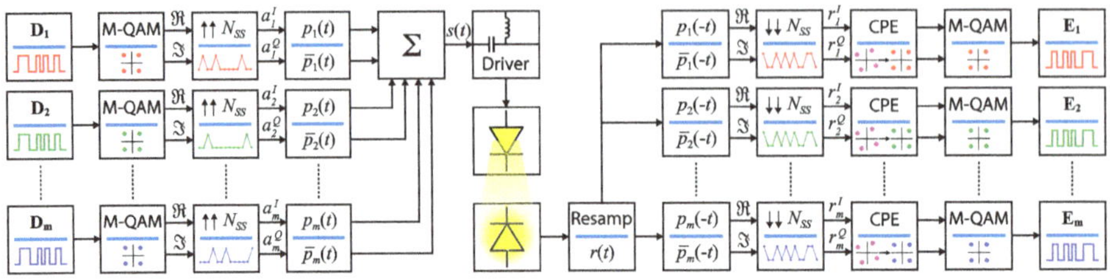

Figure 4.16. The block diagram for the super-Nyquist flavour of *m*-CAP.

$$f_{cn} = \frac{B_{sig}}{2m} - \frac{n\left[\frac{B_{sig}}{m} + B_{sig}(\alpha - 1)\right]}{m - 1} \tag{4.17}$$

recalling that B_{sig} is the total signal bandwidth and m is the total number of sub-bands. The entire receiver structure of the SNCAP system is identical to that of conventional m-CAP described previously.

The concept of SNCAP is illustrated in figure 4.17, where it is compared with conventional m-CAP (figures 4.17(a) and (b)) for $m = 2$ and for $\beta = 0.1$ and 0.5, respectively. Starting with the 2-SNCAP system, figures 4.17(c) and (d) show the ideal spectra of the proposed compressed system (blue) and the contributions of each of sub-band for $\alpha = 0.2$, i.e. 20% compression. Note that sub-band overlapping results in a 6 dB (electrical) power penalty observed at the interface. In [30] it was reported that increasing the order of m resulted in improved performance due to the alleviation of the high-frequency attention introduced by the LED [30], and hence in this work we have used up to 10 sub-bands. Note that it is expected that a 2-SNCAP VLC system would offer improved performance in comparison to higher order systems. This is because each sub-band only has a single interferer, while in higher order systems every sub-band (except the first and the last) is affected by interferers on either side. On the other hand, for high values of m and β (see figures 4.17(e) and (f)) it can be seen that; (i) the relative sub-band overlap decreases with increasing m, i.e. for 2- and 10-CAP the relative overlaps are 20% and 12.5%, respectively; and (ii) the attenuation between sub-bands actually decreases to be approximately equal to the power of the main sub-band, due to the excess bandwidth and shallower gradient of the frequency roll-off. Hence, we note that there is a trade-off between the level of IBI and the baud rate relating to the excess bandwidth factor as $R_s = B_{CAP}/(1 + \beta)$.

Obviously, the main motivation behind compressing the sub-bands is to increase the spectral efficiency η_{se}, which is given by [30]:

$$\eta_{se} = \frac{k}{(1 + \beta)(1 - \alpha)} \tag{4.18}$$

where k is the number of bits/symbol.

In [31], experimental results for the total BER as a function of m and α were measured for $\beta = \{0.1, 0.5\}$ and $k = \{2, \quad 4\}$. The total BER, as usual, is determined by the ratio of the total number of errors to the total number of bits transmitted

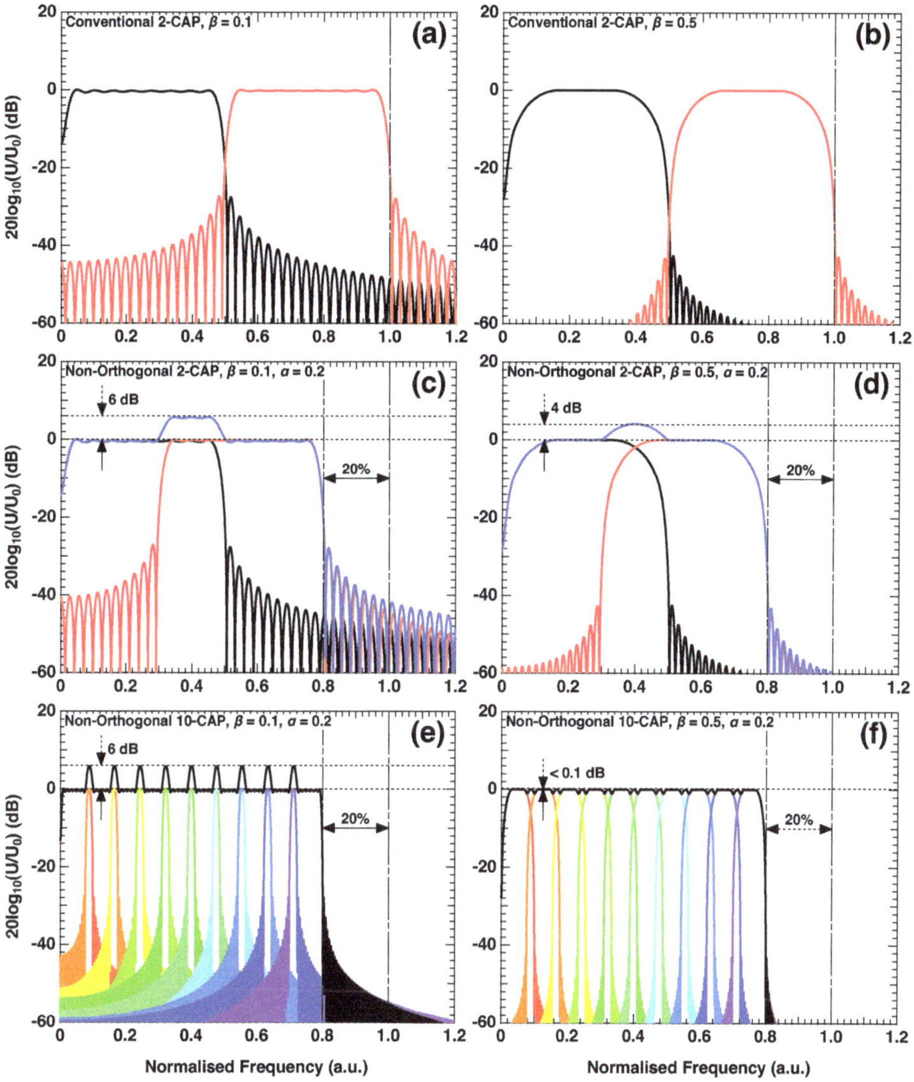

Figure 4.17. The magnitude response showing the difference between conventional CAP and SNCAP for different numbers of sub-bands and different roll-off factors β and a fixed compression; conventional CAP: (a) $\beta = 0.1$ and (b) $\beta = 0.5$, SNCAP: (c) $\beta = 0.1, \quad \alpha = 0.2$, (d) $\beta = 0.5, \quad \alpha = 0.2$, (e) $\beta = 0.1, \quad \alpha = 0.2$ and (f) $\beta = 0.5, \quad \alpha = 0.2$. Clearly, for lower values of β there is higher interference due to lower excess energy in the filters. Reproduced from [31]. CC BY 4.0.

across all sub-bands. Figure 4.18 illustrates the total measured (solid lines) and simulated (dashed lines) BER performance for the system presented in [31]. Note that the m-SNCAP BER performance improves with increasing m, particularly for $\alpha < 30\%$. This improvement is attributed to the fact that a signal of 3 MHz bandwidth is transmitted within the stop-band of the LED used, which was 1.1 MHz. Therefore the SNR improvement gained by increasing m (refer to [18])

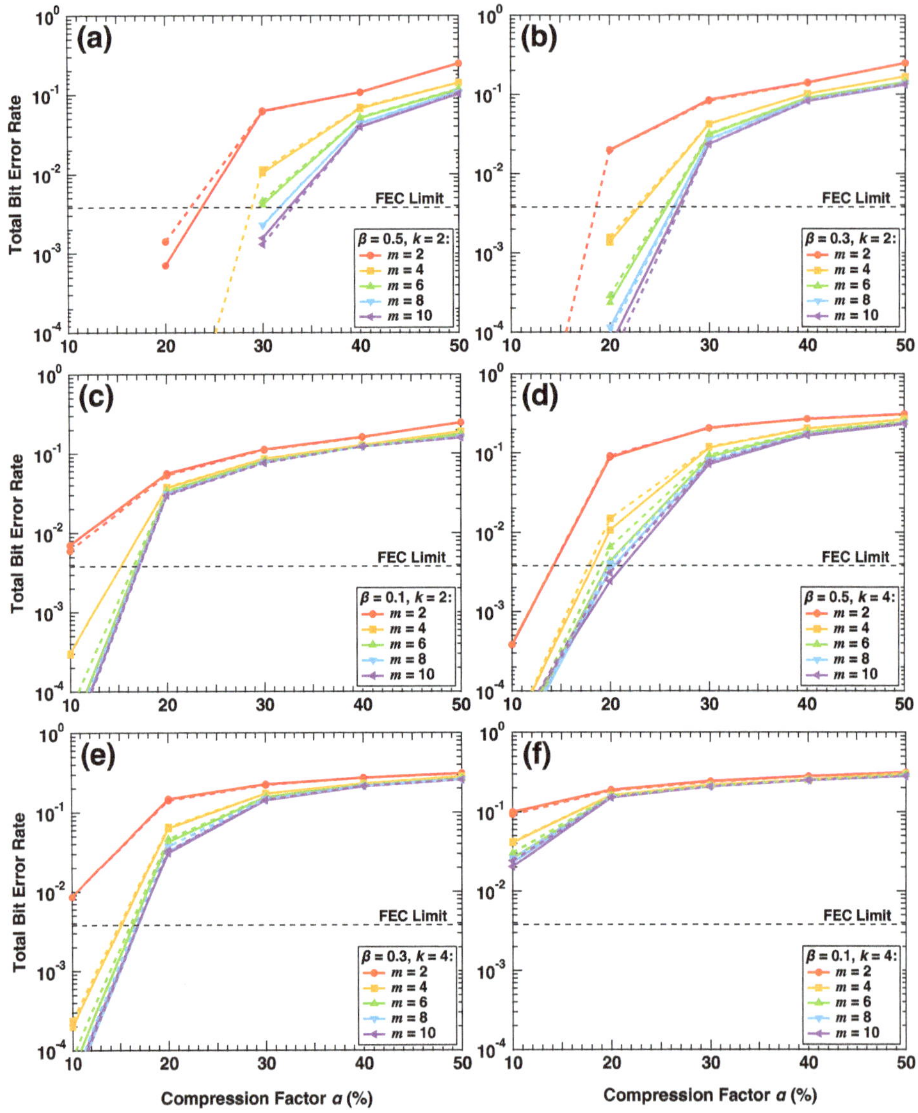

Figure 4.18. The BER performance of the SNCAP system for all sub-band orders when (a) $\beta = 0.5$, $k = 2$, (b) $\beta = 0.3$, $k = 2$, (c) $\beta = 0.1$, $k = 2$, (d) $\beta = 0.5$, $k = 4$, (e) $\beta = 0.3$, $k = 4$, (f) $\beta = 0.1$, $k = 4$. Reproduced from [31].

becomes dominant when considering the increase in IBI from more than one source when $m > 2$.

Note that higher orders of m-SNCAP (i.e. $\geqslant 8$) can support α of 30%, which corresponds to a saving in the bandwidth of 900 kHz in comparison to the uncompressed signal. Note that we measured an increase in η_{se} from 1.33 b/s/Hz in the uncompressed case using (4.18) to 1.9 b/s/Hz, for $m = 8$, $k = 2$ bits/symbol, $\alpha = 30\%$ and $\beta = 0.5$, which represents an improvement of over 40%, which is significant. This implies that the same transmission speed can be supported using a lower bandwidth, which is reduced by 30%. Therefore, the overall data rate can be increased by making use of the new, unused bandwidth for transmission of additional information. Also note that for $\alpha \geqslant 0.4$, the BER values are above the FEC limit. This is because the IBI interference is the dominant noise source and cannot be reduced using the hard threshold detection scheme, although by using advanced equalisers it may be possible.

As shown in figure 4.18(a), which offers the highest level of compression, it is clear that for $k = 2$ bits/symbol and $\beta = 0.5$ every system, i.e. any m, can support compression of at least 20% (i.e. $\alpha = 0.2$) whilst maintaining a total BER below the 7% FEC limit. The measured BER results show that the worst performance is offered by 2-SNCAP, given that each of the sub-bands only interferes with each other. This can be observed for all systems tested. The BER results for $k = 2$ and $\beta = 0.3$, see figure 4.18(b), show a reduction in a possible bandwidth compression in comparison to figure 4.18(a), since the highest compression is now $\alpha = 0.2$ for every number of sub-bands, except $m = 2$, which can only support $\alpha = 0.1$. However, since the excess bandwidth factor is reduced, η_{se} is slightly increased to 1.92 b/s/Hz, showing a marginal improvement over the previous case. Interestingly, for a given BER a lower number of sub-bands are required to achieve this value of η_{se}, that is $m = 4$ for $\beta = 0.3$, $\alpha = 0.2$ and $m = 8$ and 10 for $\beta = 0.5$ and $\alpha = 0.3$, which advantageously reduces the computational complexity by at least half [38].

As depicted in figure 4.18(c), for $\beta = 0.1$ and $k = 2$, the largest compression supported was $\alpha = 0.1$ for all m except $m = 2$, which can only support the uncompressed case at a BER value below the FEC limit. However, even though this is the smallest compression factor reported here, it corresponds to a 2.02 b/s/Hz spectral efficiency, i.e. the highest η_{se} for $k = 2$. The reason for α decreasing with β is the availability of reduced excess bandwidth, i.e. a steeper roll-off of the magnitude response of the filter and hence, a smaller inter-band guard slot and consequently the IBI dominates even for lower values of α.

Figure 4.18(d) illustrates the BER performance for $k = 4$ and $\beta = 0.5$, where $\alpha \leqslant 20\%$ can be supported to achieve a BER below the FEC limit only for $m \geqslant 8$. This indicates a clear reduction from the $k = 2$ case, where $\alpha = 0.3$ could be supported. This is clearly due to the additional SNR requirements in order to increase the modulation alphabet size from $M = 4$ to $M = 16$. Note that $\eta_{se} = 3.33$ b/s/Hz and 2.67 b/s/Hz can be supported for the system with and without compression, respectively, which corresponds to an increase of around 0.6 b/s/Hz or 25% with no additional complexity at the receiver. Figure 4.18(e) depicts the BER plots for $\beta = 0.3$ and $k = 4$. For a BER below the FEC limit, a compression factor of $\alpha = 10\%$ is supported for all m except for $m = 2$. In this case, following compression a spectral efficiency of 3.42 b/s/Hz was achieved, which is the highest value reported for $k = 4$. In comparison to the 2.67 b/s/Hz achieved for the uncompressed case, this is an improvement of 0.75 b/s/Hz (i.e. ~22%). As was the case for $k = 2$, the m-SNCAP link with $k = 4$ can be supported for $m = 4$ instead of $m > 8$ to achieve approximately the same η_{se}, while maintaining the 50% reduction in computational complexity as only half the number of finite impulse response (FIR) filters are required. Finally, for $\beta = 0.1$, no link can be supported with a BER below the FEC limit—see figure 4.18(f).

4.3 Adaptive equalisers

There are a wide variety of adaptive equalisers available and they are summed up in general form in figure 4.19 before some fundamental equaliser theory is developed. Due to the heavy band-limitation imposed by the LEDs in VLC systems, transmission often occurs outside the modulation bandwidth. In the previous discussion on modulation formats, the impact of this can be observed on CAP links (i.e. figure 4.10). The problem with transmission outside of the modulation bandwidth of the system is that ISI is introduced, which is a measure of the quantity of interference from one symbol to the following symbols, beyond the symbol duration. This effect is detrimental because the interference causes incorrect decisions when bits are recovered.

The aim of any equaliser is to remove the ISI from the system and this action is performed first by estimation of the ISI symbol span, before removal by filtering. This is best explained when considering that the frequency-domain representation of the channel response is given as $H(f)$. Given that received signals are fundamentally defined as [39]:

$$Y(f) = H(f)X(f) \tag{4.19}$$

where $Y(f)$ and $X(f)$ are the received and transmitted signal, respectively, the object of the equaliser is to generate $H_{eq}(f)$, an estimate of $H(f)$ and remove the channel as follows:

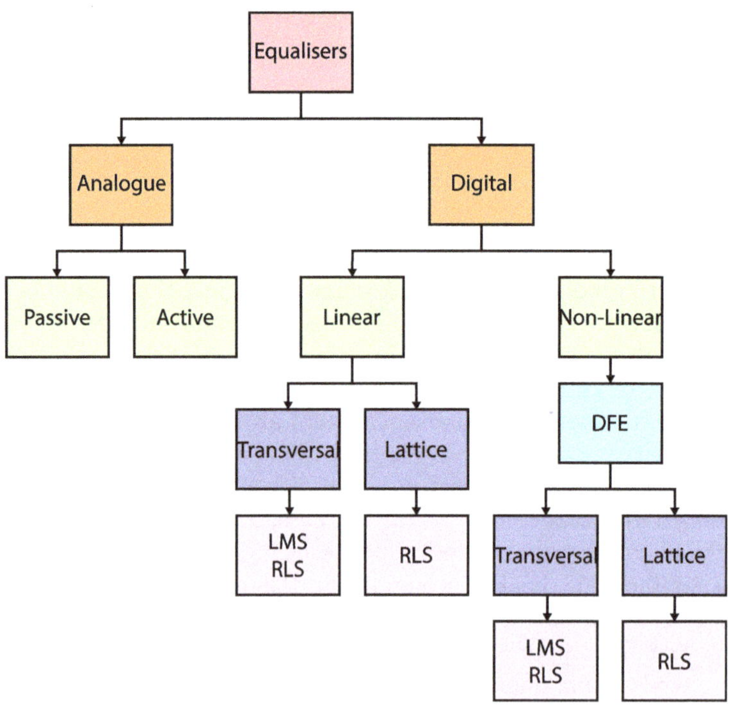

Figure 4.19. Conventional equalisation technologies.

$$Y(f) = H(f)X(f)\frac{1}{H_{eq}(f)} \tag{4.20}$$

Clearly, when $H_{eq}(f) \approx H(f)$ then the received data will be accurately represented.

One of the simplest equalisers is the linear feedforward transversal equaliser realised in FIR form. In general, the zero-forcing equaliser (ZFE) is normally introduced first as in equaliser theory, however, it is omitted here due to its well known disadvantages such as noise enhancement [40]. Linear feedforward equalisers adopt the structure shown in figure 4.20. They simply consist of several tapped delay lines (the filter length) either at the symbol rate, or a fraction of that, leading to two distinct varieties of equaliser, namely symbol and fractionally spaced. Each tapped delay line has an associated weight value, denoted w_i in figure 4.20, for the ith tap, which is trained using an adaptive algorithm. The aim of the training algorithm is to form an estimate of $H_{eq}(f) \approx H(f)$, in order to remove the ISI as outlined in (4.20). Two of the most popular algorithms to train the weight values of each input are the least mean squares (LMS) and recursive least squares (RLS) algorithms. These algorithms are well known and both work on the basis of a gradient descent on an error cost function.

4.3.1 Least mean squares

The LMS algorithm is the simpler of the two and the weights are updated using the error cost function value $\mathbf{E}\{e^2(k)\}$. It is less computationally complex than RLS because it requires no matrix inversions. The square error is given as the difference between the transmitted symbol at the kth sample $x(k)$ and estimated symbol $y(k)$ acted upon by the conjugate transpose (H) of the weights as follows [39]:

$$e^2(k) = \|x(k) - w^H y(k)\|^2 \tag{4.21}$$

The weights are updated using the value of the error cost function and the value of weight at the previous time sample as follows [39]:

$$w(k+1) = w(k) + \frac{1}{2}\mu[-\nabla\mathbf{E}\{e^2(k)\}] \tag{4.22}$$

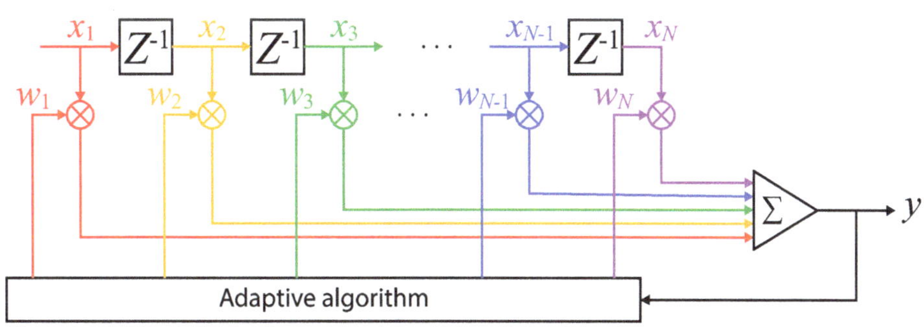

Figure 4.20. A generic architecture for transversal implementations of a feedforward equaliser.

where μ is the step-size parameter that controls the speed that the weights converge, in general a smaller value will give a more accurate convergence that takes more training time, while a larger value of μ will do the opposite. Neither of which guarantee convergence. The del operator ∇ simply indicates a gradient descent.

4.3.2 Recursive least squares

The RLS algorithm operates slightly differently because it is recursive and therefore reduces the cost function through minimisation of the linear least squares weighted error. Thus, a weighting factor $\beta_{RLS}(i, k)$ is introduced, where i is the length of the filter observation vector [39]:

$$\varepsilon(k) = \sum_{k}^{i} \beta_{RLS}(i, k) \left\| e(k) \right\|^2 \tag{4.23}$$

recalling that $e(k)$ is given in 4.21. The further particulars of both the LMS and RLS algorithms are not discussed here due to their excessive length and the fact they are well documented. For intricate details, the reader is encouraged to refer to [39].

The convergence of these two algorithms is illustrated in figures 4.21 and 4.22. Clearly, the RLS algorithm descends to a lower error magnitude at a faster convergence time if all of the previous symbols can be remembered, which is impractical. Furthermore, the RLS algorithm is more computationally complex than the LMS due to a matrix division that scales with the number of taps present in the equaliser.

4.4 Applied machine learning

While traditional equalisers such as the feedforward variety described above have been historically popular, they do not offer the best performance. Machine learning,

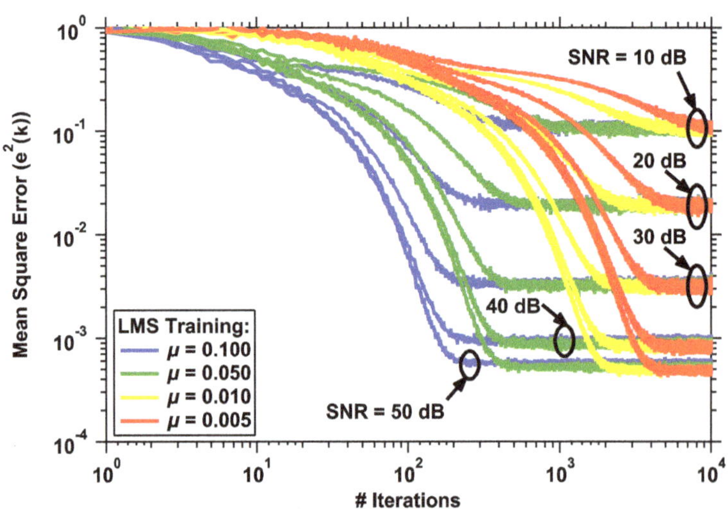

Figure 4.21. LMS convergence performance.

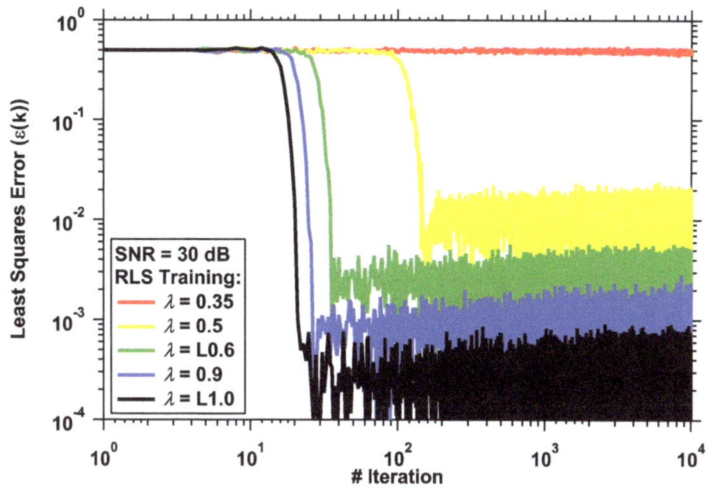

Figure 4.22. RLS convergence performance.

or artificial neural network (ANN) classifiers were first introduced by McCulloch and Pitts in 1943 [41], but popularity did not occur until decades later due to their high computational complexity that at the time simply could not be provisioned. ANNs then took a back seat until their re-emergence in the 2000s due to the availability of significantly improved computational power. A surmised history of the ANN can be found in [42, 43] and there are many new applications for ANNs across many industries including finance [44, 45], medical imaging [46, 47], pattern recognition [48, 49] and most importantly for communication systems, classification [50, 51].

One of the most common ANN applications is as an equaliser, operating by forming decision boundaries based on a training scheme. This is in opposition to calculating the contribution of ISI from each received symbol such as transversal equalisers mentioned above. The decision boundaries formed aim to classify the received symbols into groups that belong to the desired symbol value. The boundaries are formed using neurons, which can be thought of as being similar to those found in the human brain and adjust their size in reaction to the training sequence such as tap weights in transversal filters. The major difference between ANNs and transversal equalisers is their respective structures; the former are arranged into a highly parallel form that allows non-linear mapping as each input is connected to each neuron. The latter are obviously highly linear since each input is connected only to its corresponding weight.

ANNs can be divided into three distinct parts, the observation vector (input), hidden (processing) and output layers. The observation vector is an input layer that behaves akin to the tapped delay line shown in the transversal equaliser case. The hidden layer(s) is (are) where the neurons are placed and where the processing happens. The output layer is where the estimations of the bits are generated. The number of hidden layers is flexible as is the number of neurons in each layer, and they will change based on the application [52]. The ANN can be arranged in

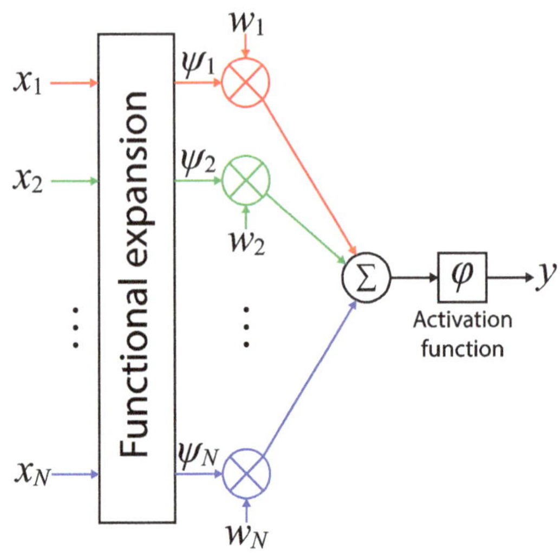

Figure 4.23. Generic architecture of a functional link ANN, including function expansion and weighted inputs to the activation function.

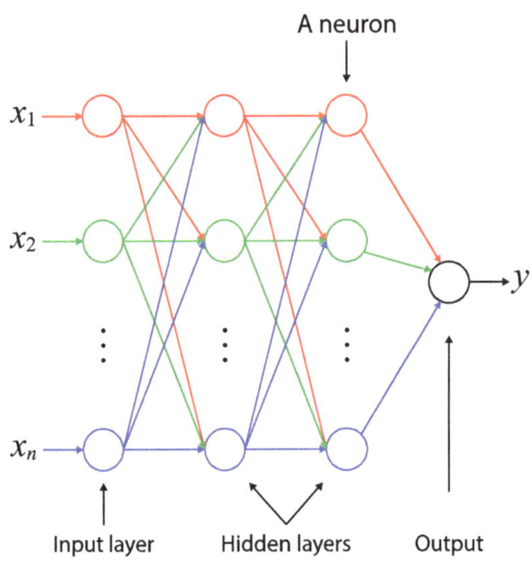

Figure 4.24. The generic MLP architecture in fully connected mode with two hidden layers.

numerous different structures and the most common include the functional link ANN and the multi-layer perceptron (MLP) illustrated respectively in figures 4.23 and 4.24. The ANN that is most commonly deployed in VLC is the fully connected MLP [53–57].

Each neuron in the MLP has a number of associated weights and the contribution from each input is scaled by the weights before summation. The schematic of a single neuron is shown in figure 4.25 along with its integration into the wider MLP

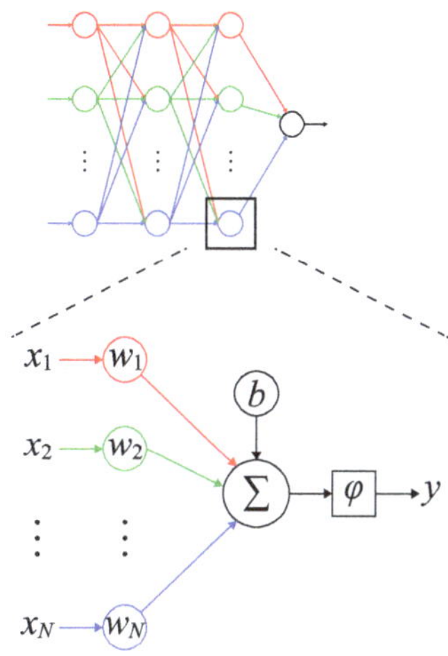

Figure 4.25. The individual architecture of a neuron including bias b.

structure. The summation can be biased using an external input. The output of a given neuron is acted on by the associated weights and biases as follows [42, 58]:

$$u(k) = b(k) + \sum_{j=1}^{N_n} y(k, j)w(k, j) \tag{4.24}$$

where there are N_n inputs to the MLP, $y(k, j)$ is the jth input to the neuron from the observation vector and $w(k, j)$ is the associated weight. The bias is given by $b(k)$. The output of the neuron is then acted upon by an activation function $\varphi(.)$, which effectively behaves as a scaling factor, restricting the possible values of the output as follows [42]:

$$z(k) = \varphi(u(k)) \tag{4.25}$$

where φ can take the form of any differentiable function, but are typically either a hard threshold, rectified linear unit (ReLU), piecewise-linear or (log-)sigmoid functions.

The log-sigmoid function is given by [42]:

$$\varphi(k) = \frac{1}{1 + \exp[-\alpha u(k)]} \tag{4.26}$$

where α is the slope parameter that controls the severity of the gradient, as can be seen in its visual representation in figure 4.26.

The most commonly employed activation function is ReLU, shown in figure 4.27, which is given by [59]:

$$\varphi(k) = \begin{cases} 0 & \text{for } u(k) < 0 \\ u(k) & \text{for } u(k) \geqslant 0 \end{cases} \tag{4.27}$$

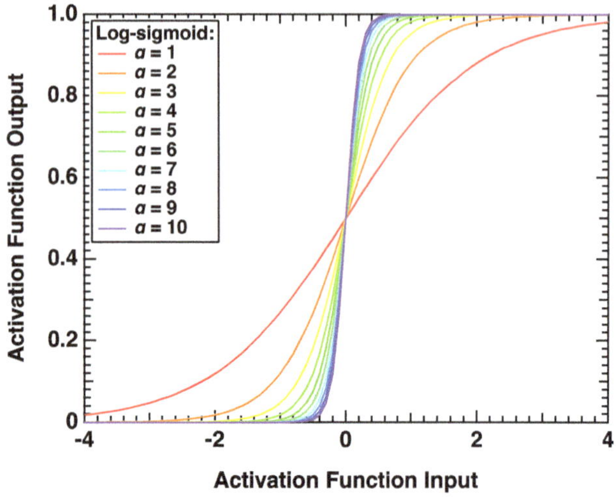

Figure 4.26. The log-sigmoid activation function with varying slope parameter *alpha*.

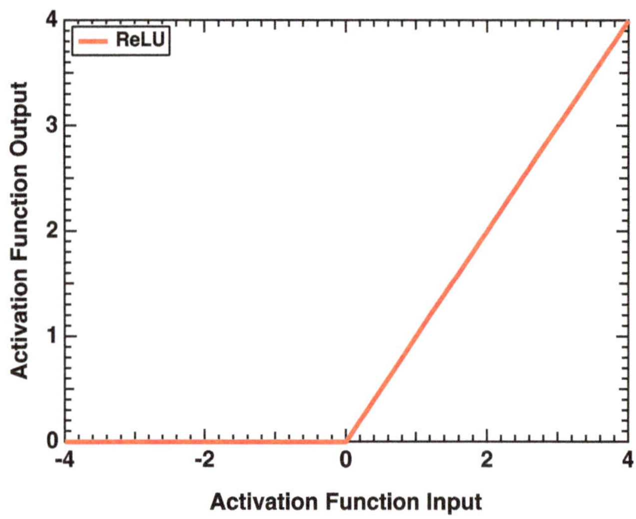

Figure 4.27. The rectified linear unit activation function.

There are numerous variants to each of these functions that have associated advantages and disadvantages, and the interested reader is encouraged to refer to [42, 43] for further information.

The output of the MLP is the summed contributions from each of the activated neurons as follows, substituting (4.24) into (4.25) [42]:

$$z(k) = \varphi\left(b(k) + \sum_{j=1}^{N_n} y(k, j)w(k, j)\right) \tag{4.28}$$

To obtain the respective weights of each neuron, the network must be properly trained. There are numerous training algorithms and the most common and popular are based on backpropagation (BP), although there are alternatives that can be found in [60]. The network is trained to minimise a loss function, which is typically the mean squared error (MSE), given by [42]:

$$E = \frac{1}{k}\sum[x(k) - z(k)]^2 \tag{4.29}$$

where $x(k)$ is the transmitted symbol and E is the MSE. The most commonly used BP algorithms are Levenberg–Marquardt (LM) and scaled conjugate gradient (SCG). The mathematics of the SCG and other BP algorithms can be found in [42, 61, 62]. Using LM-BP, the weights are calculated as follows [63]:

$$w_{ij}(k+1) = w_{ij}(k) - \eta\frac{\partial E(k)}{\partial w_{ij}(k)} \tag{4.30}$$

where w_{ij} represents the weight of the jth input to the ith neuron. The learning rate parameter η is important, since it controls the rate of MSE descent. If it is set excessively, the network will be unstable, while if it is set inadequately, the network will have a long convergence time [1, 64–66].

There have been numerous reports on the performance of MLP in VLC systems and the reader is encouraged to refer to the literature for the relative performances [67, 68]. A comparison of equalisers was made in [66], where LMS-, RLS-based transversal equalisers were compared with the MLP with LM-BP algorithm. Each system was tested with $N_n = 10$ input taps and the number of neurons is also set equivalent to N_n, with a single hidden layer. The LED used had a 2.5 MHz bandwidth, which improved to 8 MHz with a blue filter [66] at a power penalty of ~25 dB.

The relative SNR of each scheme and resulting capacity are shown in figure 4.28 and 4.29. The BER performances of each system are illustrated in figure 4.30 for the

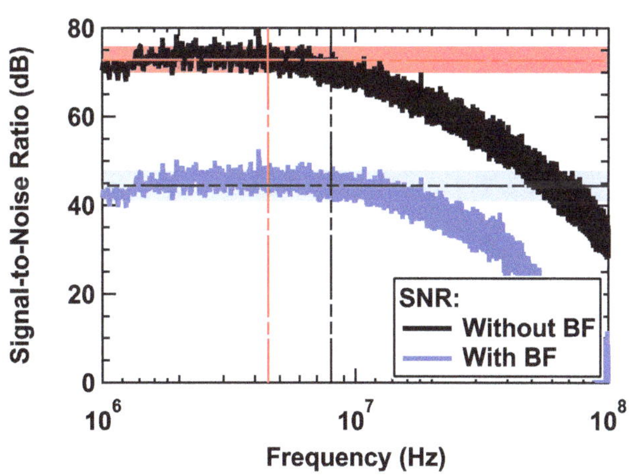

Figure 4.28. Signal-to-noise ratio as a function of frequency for the experimental system presented in [66]. © 2020 IEEE. Reprinted, with permission, from [66].

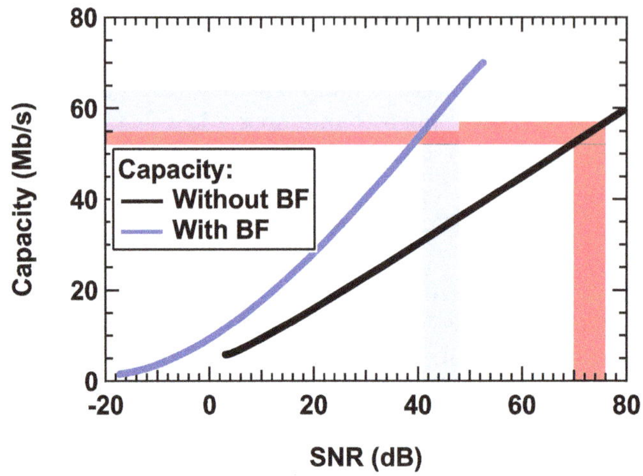

Figure 4.29. Capacity as a function of SNR for the experimental system presented in [66]. © 2020 IEEE. Reprinted, with permission, from [66].

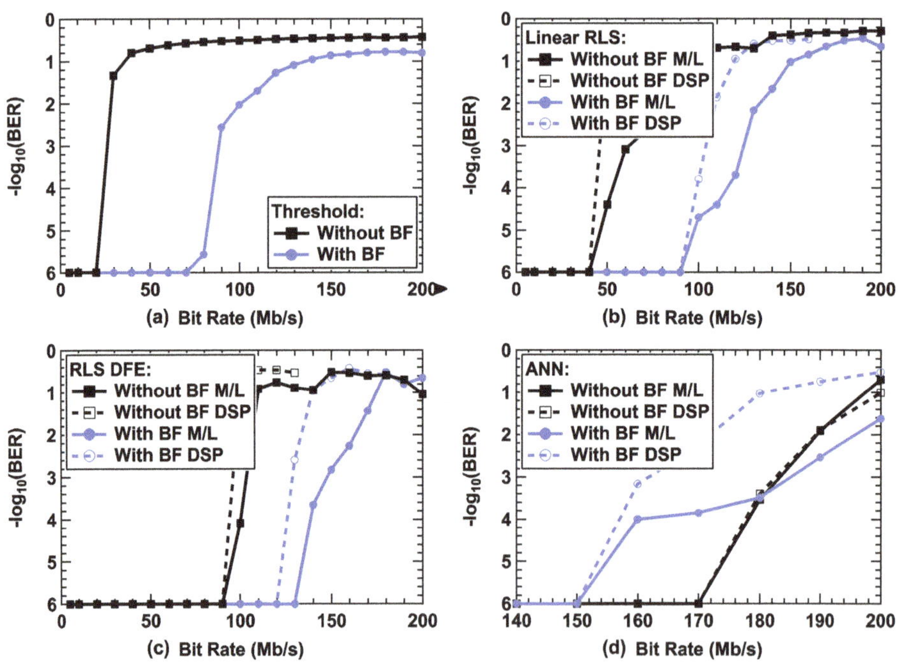

Figure 4.30. Measured BER without (black lines) and with (blue lines) blue filter for (a) the raw unequalised system presented in [66], (b) feedforward RLS, (c) decision feedback RLS and (d) an MLP-based ANN. Clearly the ANN outperforms all other systems significantly due to its superior ability to quantify the inter-symbol interference present. Interestingly, for the ANN, the white light outperforms the blue filtered system due to higher SNR. © 2020 IEEE. Reprinted, with permission, from [66].

(a) unequalised; (b) RLS feedforward; (c) RLS feedback and (d) MLP with LM training, respectively, with and without the blue filter. It was shown that the MLP substantially outperforms the conventional transversal equalisers both with and without the blue filter, achieving up to 170 Mb/s throughput.

4.5 Summary

This chapter has discussed advanced modulation formats and equalisers, including machine learning applications. In order to demonstrate high data rates for information broadcasting, squeezing the spectral efficiency of the system is of paramount importance. The most common modulation formats used such as PAM and QAM have not been covered in this chapter as they are well known. Instead, the chapter has focused on more exotic advanced modulation formats such as variations of CAP and OFDM that are currently less well known and increasing in popularity.

The chapter also discussed applied machine learning with particular focus on the MLP architecture of ANN. A comparative analysis of equalisation techniques was also presented, showing that the MLP offers superior performance due to its ability to map any input-output sequence provided there is a significant SNR.

References

[1] Ghassemlooy Z, Popoola W and Rajbhandari S 2019 *Optical Wireless Communications: System and Channel Modelling with MATLAB* (Boca Raton, FL: CRC Press)

[2] Salehi M and Proakis J 2007 Digital communications *McGraw-Hill Educ.* **31** 32

[3] Shannon C E 1948 A mathematical theory of communication *Bell Syst. Tech. J.* **27** 379–423

[4] van Nee R and Prasad R 2000 *OFDM for Wireless Multimedia Communications* (Boston, MA: Artech House)

[5] Armstrong J 2009 OFDM for optical communications *J. Light. Technol.* **27** 189–204

[6] Yang L and Armstrong J 2010 Oversampling to reduce the effect of timing jitter on high speed OFDM systems *IEEE Commun. Lett.* **14** 196–8

[7] Tubbax J, Côme B, Van der Perre L, Deneire L, Donnay S and Engels M 2001 OFDM versus single carrier with cyclic prefix: a system-based comparison *IEEE 54th Vehicular Technology Conf. VTC Fall 2001. Proc. (Cat. No. 01CH37211)* vol 2 (IEEE) pp 1115–9

[8] Raghavendra M R and Giridhar K 2004 Improving channel estimation in OFDM systems for sparse multipath channels *IEEE Signal Process. Lett.* **12** 52–5

[9] Rodrigues M and Darwazeh I 2002 Fast OFDM: a proposal for doubling the data rate of OFDM schemes *Proc. Int. Conf. on Telecommunications* vol 3 pp 484–7

[10] Ahmed N, Natarajan T and Rao K R 1974 Discrete cosine transform *IEEE Trans. Comput.* **100** 90–3

[11] Rao K R and Yip P 2014 *Discrete Cosine Transform: Algorithms, Advantages, Applications* (New York: Academic)

[12] Haigh P A and Darwazeh I 2017 Visible light communications: fast-orthogonal frequency division multiplexing in highly bandlimited conditions *2017 IEEE/CIC Int. Conf. on Communications in China (ICCC Workshops)* pp 1–8

[13] Olmedo M I, Zuo T, Jensen J B, Zhong Q, Xu X, Popov S and Monroy I T 2013 Multiband carrierless amplitude phase modulation for high capacity optical data links *J. Lightwave Technol.* **32** 798–804

[14] Landau H J 1967 Sampling, data transmission, and the Nyquist rate *Proc. IEEE* **55** 1701–6

[15] Haigh P A 2019 Carrier-less amplitude and phase modulation in visible light communications: recent progress *2019 6th NAFOSTED Conf. on Information and Computer Science (NICS)* (IEEE) pp 109–14

[16] Haigh P A, Chvojka P, Minotto A, Burton A, Murto P, Wang E, Ghassemlooy Z, Zvanovec S, Cacialli F and Darwazeh I 2019 Hybrid super-Nyquist CAP modulation based VLC with low bandwidth polymer LEDs *2019 IEEE 30th Annual Int. Symp. on Personal, Indoor and Mobile Radio Communications (PIMRC)* (IEEE) pp 1–6

[17] Haigh P A, Chvojka P, Zvanovec S, Ghassemlooy Z and Darwazeh I 2018 Analysis of Nyquist pulse shapes for carrierless amplitude and phase modulation in visible light communications *J. Lightwave Technol.* **36** 5023–9

[18] Haigh P A, Burton A, Werfli K, Minh H L, Bentley E, Chvojka P, Popoola W O, Papakonstantinou I and Zvanovec S 2015 A multi-CAP visible-light communications system with 4.85-b/s/Hz spectral efficiency *IEEE J Select. Areas Commun.* **33** 1771–9

[19] Chvojka P, Werfli K, Zvanovec S, Haigh P A, Vacek V H, Dvorak P, Pesek P and Ghassemlooy Z 2017 On the *m*-CAP performance with different pulse shaping filters parameters for visible light communications *IEEE Photonics J.* **9** 1–12

[20] Chvojka P, Zvanovec S, Werfli K, Haigh P A and Ghassemlooy Z 2017 Variable *m*-CAP for bandlimited visible light communications *IEEE Int. Conf. on Communications Workshops, ICC Workshops 2017* (IEEE) pp 1–5

[21] Werfli K, Haigh P A, Ghassemlooy Z, Hassan N B and Zvanovec S 2016 A new concept of multi-band carrier-less amplitude and phase modulation for bandlimited visible light communications *2016 10th Int. Symp. on Communication Systems, Networks and Digital Signal Processing (CSNDSP)* (IEEE) pp 1–5

[22] Bykhovsky D and Arnon S 2014 An experimental comparison of different bit-and-power-allocation algorithms for DCO-OFDM *J. Lightwave Technol.* **32** 1559–64

[23] Stepniak G 2018 Staggered CAP: a new spectrally efficient modulation format for optical communications *IEEE Photonics Technol. Lett.* **30** 367–70

[24] Lin B, Tang X, Ghassemlooy Z, Fang X, Lin C, Li Y and Zhang S 2016 Experimental demonstration of OFDM/OQAM transmission for visible light communications *IEEE Photonics J.* **8** 1–10

[25] Zhao J and Ellis A D 2011 Offset-QAM based coherent WDM for spectral efficiency enhancement *Opt. Express* **19** 14617–31

[26] Haigh P A *et al* 2019 Experimental demonstration of staggered CAP modulation for low bandwidth red-emitting polymer-LED based visible light communications *2019 IEEE Int. Conf. on Communications Workshops (ICC Workshops)* (IEEE) pp 1–6

[27] Akande K O and Popoola W O 2018 Subband index carrierless amplitude and phase modulation for optical communications *J. Lightwave Technol.* **36** 4190–7

[28] Wang Y, Tao L, Wang Y and Chi N 2014 High speed wdm VLC system based on multi-band CAP-64 with weighted pre-equalization and modified CMMA based post-equalization *IEEE Commun. Lett.* **18** 1719–22

[29] Liang S, Qiao L, Lu X and Chi N 2018 Enhanced performance of a multiband super-Nyquist CAP-16 VLC system employing a joint MIMO equalizer *Opt. Express* **26** 15718–25
[30] Haigh P A, Chvojka P, Ghassemlooy Z, Zvanovec S and Darwazeh I 2018 Non-orthogonal multi-band CAP for highly spectrally efficient VLC systems *11th Int. Symp. on Communication Systems, Networks & Digital Signal Processing (CSNDSP)* pp 1–6
[31] Haigh P A, Chvojka P, Ghassemlooy Z, Zvanovec S and Darwazeh I 2019 Visible light communications: multi-band super-Nyquist CAP modulation *Opt. Express* **27** 8912–9
[32] Yu J J, Zhang J W, Dong Z, Jia Z S, Chien H C, Cai Y, Xiao X and Li X Y 2013 Transmission of 8 × 480-Gb/s super-Nyquist-filtering 9-QAM-like signal at 100 GHz-grid over 5000-km SMF-28 and twenty-five 100 GHz-grid ROADMs *Opt. Express* **21** 15686–91
[33] Anderson J B, Rusek F and Owall V 2013 Faster-than-Nyquist signaling *Proc. IEEE* **101** 1817–30
[34] Darwazeh I, Xu T Y, Gui T, Bao Y and Li Z H 2014 Optical SEFDM system; bandwidth saving using non-orthogonal sub-carriers *IEEE Photonics Technol. Lett.* **26** 352–5
[35] Nopchinda D, Xu T Y, Maher R, Thomsen B C and Darwazeh I 2016 Dual polarization coherent optical spectrally efficient frequency division multiplexing *IEEE Photonics Technol. Lett.* **28** 83–6
[36] Xu T, Mikroulis S, Mitchell J E and Darwazeh I 2016 Bandwidth compressed waveform for 60-GHz millimeter-wave radio over fiber experiment *J. Lightwave Technol.* **34** 3458–65
[37] Darwazeh I, Ghannam H and Xu T 2018 The first 15 years of SEFDM: A brief survey *11th Int. Symp. on Communication Systems, Networks and Digital Signal Processing (CSNDSP)* (IEEE) pp 1–5
[38] Wei J, Sanchez C, Haigh P A and Giacoumidis E 2017 Complexity comparison of multi-band CAP and DMT for practical high speed data center interconnects *Asia Communications and Photonics Conf.* (Optical Society of America) p M2G.4
[39] Haykin S S 2005 *Adaptive Filter Theory* (New Delhi: Pearson Education India)
[40] Chen C-J and Wang L-C 2007 Performance analysis of scheduling in multiuser MIMO systems with zero-forcing receivers *IEEE J. Sel. Areas Commun.* **25** 1435–45
[41] McCulloch W S and Pitts W 1943 A logical calculus of the ideas immanent in nervous activity *Bull. Math. Biophys.* **5** 115–33
[42] Haykin S 2010 *Neural Networks and Learning Machines* 3rd edn (New Delhi: Pearson Education India)
[43] Haykin S 2004 A comprehensive foundation *Neural Netw.* **2** 41
[44] Kuo R J, Chen C H and Hwang Y C 2001 An intelligent stock trading decision support system through integration of genetic algorithm based fuzzy neural network and artificial neural network *Fuzzy Sets Syst.* **118** 21–45
[45] Chang P-C, Liu C-H, Lin J-L, Fan C-Y and Ng C S P 2009 A neural network with a case based dynamic window for stock trading prediction *Expert Syst. Appl.* **36** 6889–98
[46] Chen H, Zhang Y, Kalra M K, Lin F, Chen Y, Liao P, Zhou J and Wang G 2017 Low-dose CT with a residual encoder-decoder convolutional neural network *IEEE Trans. Med. Imaging* **36** 2524–35
[47] Lee J-G, Jun S, Cho Y-W, Lee H, Kim G B, Seo J B and Kim N 2017 Deep learning in medical imaging: general overview *Korean J. Radiol.* **18** 570–84
[48] Pao Y 1989 *Adaptive Pattern Recognition and Neural Networks* (New York: Addison-Wesley)

[49] Carpenter G A and Grossberg S 1988 The art of adaptive pattern recognition by a self-organizing neural network *Computer* **21** 77–88
[50] Dreiseitl S and Ohno-Machado L 2002 Logistic regression and artificial neural network classification models: a methodology review *J. Biomed. Inform.* **35** 352–9
[51] Hepner G, Logan T, Ritter N and Bryant N 1990 Artificial neural network classification using a minimal training set-comparison to conventional supervised classification *Photogramm. Eng. Remote Sens.* **56** 469–73
[52] Karsoliya S 2012 Approximating number of hidden layer neurons in multiple hidden layer BPNN architecture *Int. J. Eng. Trends Technol.* **3** 714–7
[53] Rajbhandari S, Chun H, Faulkner G, Haas H, Xie E, McKendry J J D, Herrnsdorf J, Gu E, Dawson M D and O'Brien D 2019 Neural network-based joint spatial and temporal equalization for MIMO-VLC system *IEEE Photonics Technol. Lett.* **31** 821–4
[54] Osahon I N, Pikasis E, Rajbhandari S and Popoola W O 2017 Hybrid POF/VLC link with *M*-PAM and MLP equaliser *2017 IEEE Int. Conf. on Communications (ICC)* (IEEE) pp 1–6
[55] Haigh P A *et al* 2014 A 20-Mb/s VLC link with a polymer LED and a multilayer perceptron equalizer *IEEE Photonics Technol. Lett.* **26** 1975–8
[56] Rajbhandari S, Haigh P A, Ghassemlooy Z and Popoola W 2013 Wavelet-neural network VLC receiver in the presence of artificial light interference *IEEE Photonics Technol. Lett.* **25** 1424–7
[57] Li G, Hu F, Zhao Y and Chi N 2019 Enhanced performance of a phosphorescent white LED CAP 64QAM VLC system utilizing deep neural network (DNN) post equalization *2019 IEEE/CIC Int. Conf. on Communications in China (ICCC)* (IEEE) pp 173–6
[58] Cambria E *et al* 2013 Extreme learning machines (trends and controversies) *IEEE Intell. Syst.* **28** 30–59
[59] Xu B, Wang N, Chen T and Li M 2015 Empirical evaluation of rectified activations in convolutional network arXiv:1505.00853
[60] Rios L M and Sahinidis N V 2013 Derivative-free optimization: a review of algorithms and comparison of software implementations *J. Glob. Optim.* **56** 1247–93
[61] Chauvin Y and Rumelhart D E 1995 *Backpropagation: Theory, Architectures, and Applications* (London: Psychology Press)
[62] Rumelhart D E, Durbin R, Golden R and Chauvin Y 1995 Backpropagation: the basic theory *Backpropagation: Theory, Architectures and Applications* (London: Psychology Press) pp 1–34
[63] Moré J J 1978 The Levenberg–Marquardt algorithm: implementation and theory *Numerical Analysis* (Berlin: Springer) pp 105–16
[64] Haigh P A, Ghassemlooy Z, Le Minh H, Rajbhandari S, Arca F, Tedde S F, Hayden O and Papakonstantinou I 2012 Exploiting equalization techniques for improving data rates in organic optoelectronic devices for visible light communications *J. Lightwave Technol.* **30** 3081–8
[65] Burse K, Yadav R N and Shrivastava S C 2010 Channel equalization using neural networks: a review *IEEE Trans. Syst. Man Cybern. C (Appl. Rev.)* **40** 352–7
[66] Haigh P A, Ghassemlooy Z, Rajbhandari S, Papakonstantinou I and Popoola W 2014 Visible light communications: 170 Mb/s using an artificial neural network equalizer in a low bandwidth white light configuration *J. Lightwave Technol.* **32** 1807–13

[67] Rajbhandari S, Faith J, Ghassemlooy Z and Angelova M 2013 Comparative study of classifiers to mitigate intersymbol interference in diffuse indoor optical wireless communication links *Optik* **124** 4192–6
[68] Osahon I N, Rajbhandari S and Popoola W O 2018 Performance comparison of equalization techniques for SI-POF multi-gigabit communication with PAM-*M* and device non-linearities *J. Lightwave Technol.* **36** 2301–8

IOP Publishing

Visible Light
Data communications and applications
Paul Anthony Haigh

Chapter 5

Balancing lighting with data communications

5.1 Introduction

As VLC systems are presented as dual-function, dual benefit, it is important that the user has total control over the illumination and understands the trade-offs between the received SNR, power and available transmission rates. Therefore, controlling the level of light output in VLC systems is of the utmost importance. One must ensure the end-user has complete control over the illumination level on the receiving plane, particularly in a home and office environment where there are different scenarios to consider.

The end-user may wish to dim the lights to watch a film streamed through their VLC system and as such, the information rate that is supplying the underlying data must maintain a minimum level of performance. There have been myriad extensive studies and proposals in VLC systems on dimming in combination with modulation, and more recently, looking at colour balancing of the wavelengths to maintain a flicker-free, white-balanced link. Chapter 5 provides a review of the latest developments in this field.

5.2 Dimming in visible light communication

First, a discussion on the impact of dimming in a VLC system will be presented. The human eye perceives light with the square root of that measured by a photodiode, as reported by the Illumination Engineering Society of North America [1]:

$$\text{Perceived light (\%)} = 100 \times \sqrt{\frac{\text{Measured light (\%)}}{100}} \tag{5.1}$$

where the factors of 100 are included to maintain the percentage values. Equation (5.1) is illustrated in figure 5.1 to highlight the fact that the human eye has a non-linear response to dimming, and hence high resolution dimming control is required in order to give the end-user the desired performance. For instance, from figure 5.1,

doi:10.1088/978-0-7503-1680-4ch5

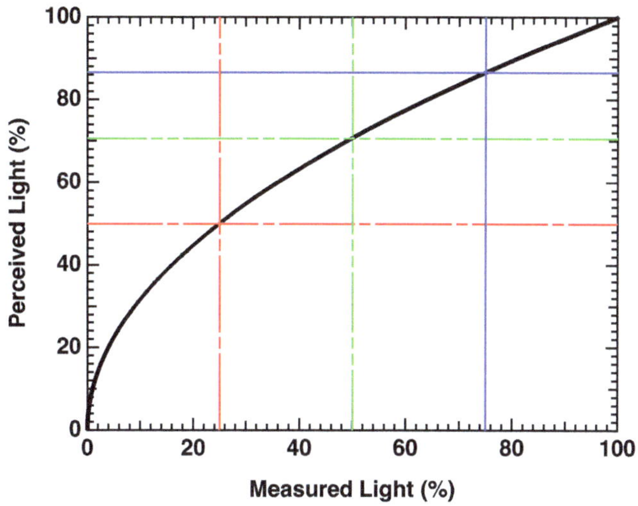

Figure 5.1. The relationship between measured light using a power meter and perceived light by the human eye, which is clearly non-linear. For instance, the absolute value of 50% measured light translates to just over 70% of the same power by the human eye.

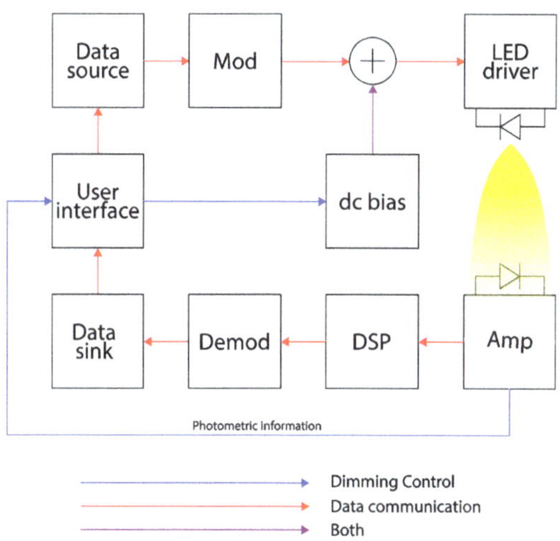

Figure 5.2. A generic block diagram for a dimming system, which differs from previous systems due to the addition of the user interface that controls the dc bias and/or other elements of the system.

there are three cases highlighted, when a lamp is dimmed to 25%, 50% and 75% of the maximum, the perceived light is estimated by the human eye as 50%, ~71% and ~87%, respectively.

The general block diagram for a dimming system is presented in figure 5.2, adopted from [2]. Furthermore, considering the human eye sensitivity function $V(\lambda)$, the luminous flux Φ impinging on the LED is given by [3]:

$$\Phi = 683\left[\frac{\text{lm}}{\text{W}}\right]\int_{380\text{nm}}^{780\text{nm}} P_R(\lambda)V(\lambda)\ \text{d}\lambda \tag{5.2}$$

where $P_R(\lambda)$ is the received power at the wavelength λ. The illuminance E_v is a function of the luminous flux and is given by [4]:

$$E_\text{v} = \frac{\Phi}{A_\text{pd}} \tag{5.3}$$

where A_pd is the area of the PD receiving the light. These relationships are useful to gain an idea of the light levels required as values are generally given in lux. The human eye sensitivity function is illustrated in 5.1 [5].

Typical values of illuminance in different environments are stated in table 5.1 [5]. Dimming is achieved by varying the average light level and is relative to the optical power level as follows [6]:

$$x(t) \geqslant 0 \text{ and } \lim_{T\to\infty}\frac{1}{T}\int_0^T x(t)\ \text{d}t \geqslant \gamma P_\text{avg} \tag{5.4}$$

where P_avg is the average power and $0 \leqslant \gamma \leqslant 1$ is the dimming factor. There are generally two key physical methods that can be used to control the light level of the LED. The first one is to vary the dc current that drives the LED where the light level follows the current level, within the linear operating region of the device. This method is colloquially known as continuous current reduction (CCR) and its key advantage is that it's trivial to control the drive current of the LED in either of the analogue or digital domains. The main disadvantage, however, is that the absolute optical power output will drop with the signal level and therefore the received SNR will also be reduced, which in turn reduces the capacity, following Shannon [7]. Furthermore, reducing the dc drive current also decreases the modulation bandwidth [8] and shifts the wavelength emitted [9]. This can be a huge issue if the system is optimised to operate at a nominal bandwidth or wavelength.

The other method generally employed is based on the manipulation of digital modulation schemes. In the literature, numerous approaches have taken this approach to maintain data rates and provide control on the light level. Clearly, it is impossible to maintain the total throughput whilst reducing the signal power

Table 5.1. Typical values of illuminance for common lighting scenarios.

Illuminance condition	Illuminance (lux)
Full Moon	1
Street lighting	10
Home lighting	30–300
Office lighting	100–1000
Surgery lighting	10 000
Direct sunlight	100 000

(i.e. reducing the SNR) or introducing 'off' time into the signal (puncturing the rate). One of the first and most common approaches has been PWM, since it is straightforward to control the mark-space ratio to control the average light level [10–13]. Others have proposed PPM for similar reasons, as the energy distribution of this format is easy to control and thus the light level can also be controlled with ease [14–16]. Furthermore, subcarrier index modulation (SIM) has also become popular for dimming control in recent years [17–20]. In SIM, which takes advantage of multi-carrier modulation, certain subcarriers or sub-bands are dynamically switched on or off depending on the required average power. This section will focus on these three key approaches.

5.2.1 Pulse width modulation

The first approach discussed is PWM, which is uncommonly used in modern data communication systems, but the basic principles can be referred to in [21]. The fundamental reason why researchers have investigated PWM is because it is extremely simple to implement, as is highlighted in figure 5.3. A PWM system has a mark-space ratio which simply means to divide the symbol period into an 'on' (mark) and 'off' (space) period in a repetitive manner, and the summed duration of both is equivalent to the symbol period. This is given mathematically as follows [22]:

$$x_{\mathrm{PWM}}(t) = \begin{cases} x_{\mathrm{H}}(t) & \text{for } 0 < t < T_{\mathrm{M}} \\ x_{\mathrm{L}}(t) & \text{for } T_{\mathrm{M}} < t \leqslant T_{\mathrm{PWM}} \end{cases} \tag{5.5}$$

where $x_{\mathrm{PWM}}(t)$, $x_{\mathrm{H}}(t)$, $x_{\mathrm{L}}(t)$ are the overall PWM signal, the mark amplitude and the space amplitude, respectively, t is the instantaneous time, T_H is the mark duration and T_{PWM} is the overall signal duration. The space duration T_S is clearly given by $T_{\mathrm{S}} = T_{\mathrm{PWM}} - T_{\mathrm{H}}$.

Increasing the mark duration (decreasing the space duration) increases the average signal power and therefore the amount of light collected at the receiver.

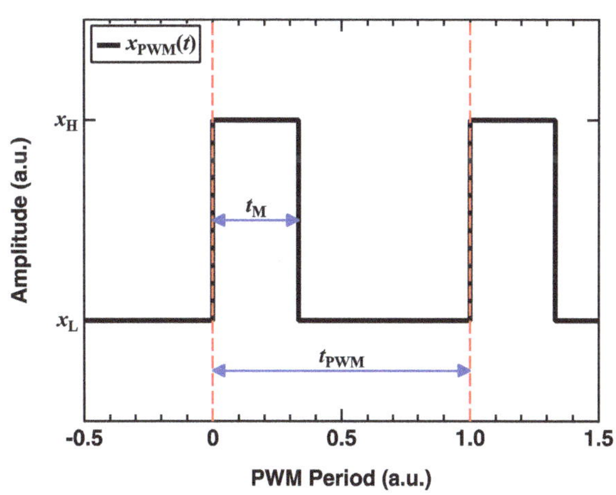

Figure 5.3. The PWM concept.

Similarly, doing the opposite and decreasing the mark period decreases the average power and therefore reduces the light level at the receiver.

Generally the first propositions of achieving dimming using PWM were to superpose a simple modulation format such as OOK onto the PWM mark duration as follows [22]:

$$x_{\text{signal}}(t)\begin{cases} x_{\text{OOK}}(t)x_{\text{H}}(t) & \text{for } 0 < t < T_{\text{M}} \\ 0 & \text{for } T_{\text{M}} < t \leqslant T_{\text{PWM}} \end{cases} \tag{5.6}$$

where $x_{\text{OOK}}(t)$ is the OOK signal containing data; effectively (5.6) means that data can be transmitted in the 'on' period, adopting the amplitude of the PWM signal, but remains inactive when the system is 'off'. Using the approach resulted in a throughput that was directly proportional to the mark-space ratio and therefore, the average light level. Clearly inefficient, the next natural step was to superpose data onto both the mark and space durations, albeit with different amplitudes for both, as follows [23]:

$$x_{\text{signal}}(t) = \begin{cases} x_{\text{OOK}}(t)x_{\text{H}}(t) & \text{for } 0 < t < T_{\text{M}} \\ x_{\text{OOK}}(t)x_{\text{L}}(t) & \text{for } T_{\text{M}} < t \leqslant T_{\text{PWM}} \end{cases} \tag{5.7}$$

which holds and introduces an increase in data rate provided $x_{\text{L}}(t) > 0$. An alternative method, which requires more bandwidth, is to adapt the rate to compensate for the outage time. These concepts are all illustrated for clarity in figure 5.4.

The idea of superposing different modulation formats onto a PWM signal was eventually extended to high spectral efficiency modulation formats such as OFDM. There are two generic flavours of OFDM, including the dc-biased OFDM described in chapter 4. The other main configuration is called asymmetrically-clipped OFDM and which produces a positive signal by only modulating the odd subcarriers and clipping any value less than zero as follows [24]:

$$x(t) = \begin{matrix} x_0(t) \text{ if } x_0(t) > 0 \\ 0 \text{ if } x_0(t) \leqslant 0 \end{matrix} \tag{5.8}$$

where $x_0(t)$ is the original bipolar OFDM signal. The main issue with this method is that when considering the Hermitian symmetry requirements, only $N/4$ of the subcarriers are loaded, where N is the total number of subcarriers used, since only the odd subcarriers are loaded. The reason only the odd subcarriers are loaded is because it has been demonstrated in the literature that the inter-modulation between subcarriers only occurs in the locations of the even subcarriers after clipping [25]. The generation of the asymmetrically-clipped signal is given as [25]. The generation of the asymmetrically-clipped signal is as given in (4.2), however, the data modulating the subcarriers is $\mathbf{X} = [0, X_1, 0, X_3, \cdots, X_{N-1}]$. An example of an asymmetrically-clipped OFDM signal is shown in figure 5.5.

This asymmetrically-clipped technique may not be the most spectrally efficient modulation format available, but it offers several key advantages over OOK, and

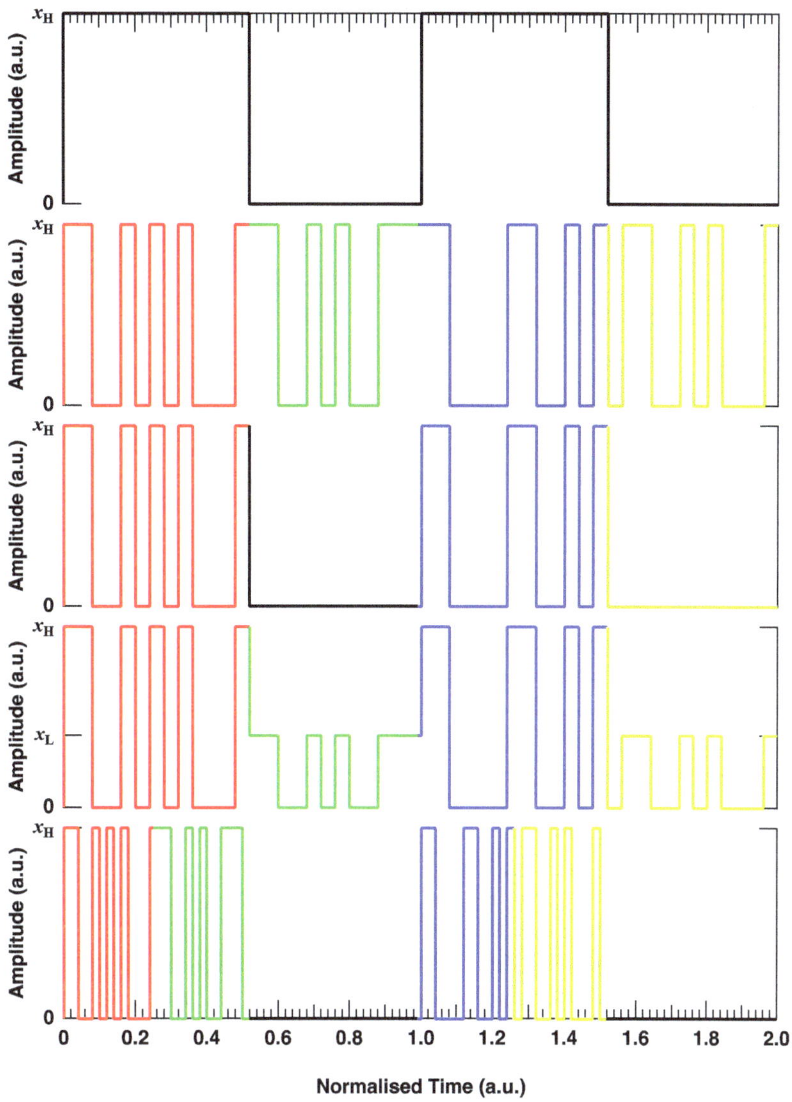

Figure 5.4. Different methods of achieving dimming. The top graph shows the generic PWM square wave; the second is the data to be transmitted; the third is a punctured approach to PWM, where data is lost when the PWM signal is off; the fourth is an amplitude adaptive system, where in the PWM 'off' time, the amplitude of the signal is significantly reduced to maintain data rate; and finally a rate adaptive approach where the data rate is increased to compensate for the 'off'-period.

the most important is being able to load the different subcarriers with different numbers of bits-per-symbol, which leads to higher aggregated efficiencies and adaptation to the received SNR. This technique was introduced in [22] where asymmetrically-clipped OFDM symbols are generated and the polarity of a portion of the signal is negated according to the PWM signal properties. This was named reverse polarity optical OFDM and is mathematically defined as [22]:

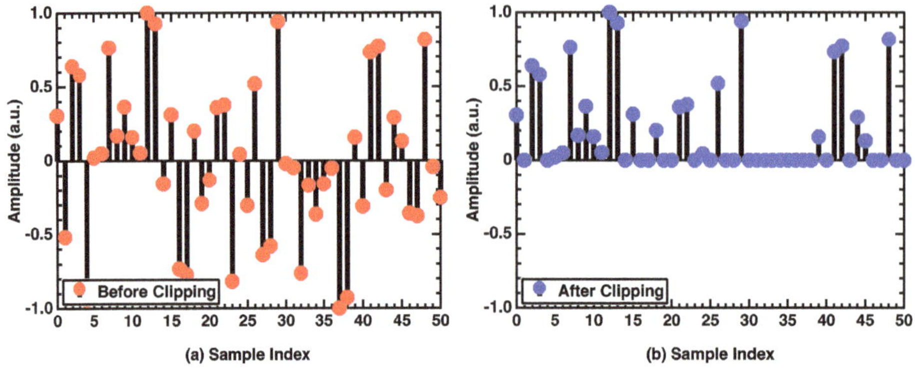

Figure 5.5. An example of an asymmetrically-clipped OFDM signal, before and after clipping.

$$x(t) = \begin{cases} x_{\mathrm{H}}(t) - mx_{\mathrm{OFDM}}(t) & \text{for } 0 < t < T_{\mathrm{M}} \\ x_{\mathrm{H}}(t) + mx_{\mathrm{OFDM}}(t) & \text{for } T_{\mathrm{M}} < t \leqslant T_{\mathrm{PWM}} \end{cases} \tag{5.9}$$

where m is simply a scaling factor and x_{OFDM} is the reverse polarity OFDM signal, as shown in figure 5.6 for several duty cycles. This approach was shown to maintain data rates effectively whilst providing dual control on the dimming level. The dual control emerges from the mark-space ratio and also the value of m, which can be controlled to tweak the signal level as desired. The authors show that dimming levels down to 5% can be supported at sufficiently low BERs.

5.2.2 Pulse position modulation

The second method to be discussed is PPM, which is effectively a modulation format that broadly divides the symbol period into slots, where traditionally only a single slot contains energy. This makes it a highly power efficient modulation when a higher number of slots are selected. The first text to seriously discuss PPM as an approach to dimming was [6], which proposed modifications to both PPM and OOK to transform them into variable schemes with dimming control. In general PPM, the number of slots is defined as L and the slot duration $T_L = T/L$. The PPM order is generally set to powers of two for compatibility with the binary nature of the data used. The overall L-PPM signal is given by [6]:

$$x(t) = LP_{\mathrm{avg}} \sum_{l}^{L-1} \mathbf{C}_l g(t - lT_L) \tag{5.10}$$

where $g(.)$ is the pulse shaping filter used, typically rectangular, and $\mathbf{C}_l$ is the PPM codeword that typically only contains a single 'one' value in any given slot, $\mathbf{C}_l = [c_0, c_1, \cdots, c_{L-1}]$. The main limitation for variable pulse position modulation (VPPM), however, is that it is generally set to $L = 2$ slots.

The reason for the restriction that $L = 2$ is because the slot width is varied as a function of the required average brightness. This leads to the waveform generated in figure 5.7 that shows the equivalent data symbols to transmit for a PPM signal, and

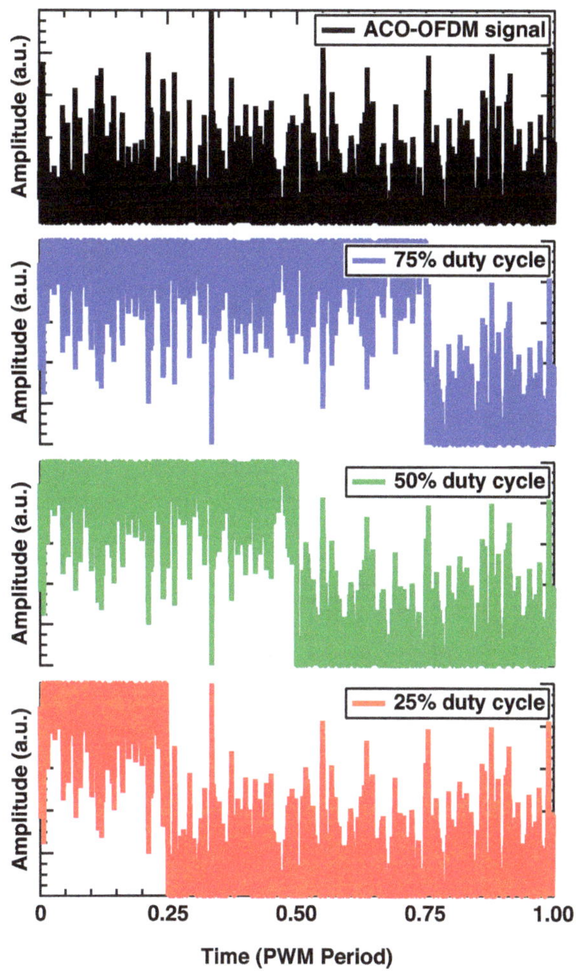

Figure 5.6. The reverse polarity system proposed in [22] for a number of different duty cycles.

also VPPM signals for 25%, 50% and 75% slot widths. Clearly, when the slot width is 50%, it is equivalent to traditional PPM. Considering that L is set as a power of two and the objective of varying the modulation format is to control dimming, the easiest way to ensure that 100% and 0% dimming can be achieved with ease is to set $L = 2$, however the trade-off for these two extremes is that a dc level is transmitted and no information is transmitted. If $L > 2$ and keeping consistent with the methodology of PPM where only a single slot can contain energy, the maximum brightness that could be achieved falls away with L.

In early VLC standards, VPPM was adopted, i.e. in [15]; however, as mentioned above, the PPM variants have considerable drawbacks and have limited research value, in the opinion of the author.

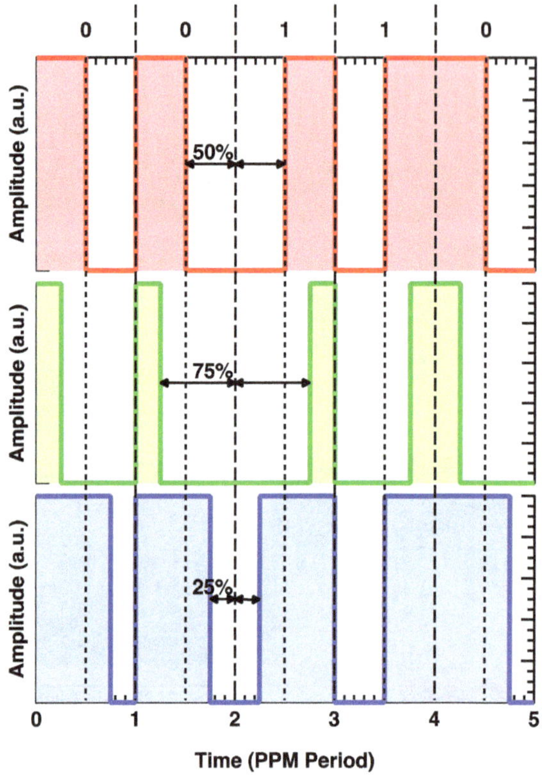

Figure 5.7. The concept of VPPM where the pulse is broadened or shortened depending on the desired dimming level.

5.2.3 Subcarrier index modulation

One of the more innovative approaches to dimming has been the introduction of SIM to VLC systems, which is an emerging technique that acts as a second order modulation, on top of a traditional multi-carrier scheme such as OFDM or *m*CAP, which improves spectral efficiency by encoding further data into the system and also energy efficiency by reduction of the peak-to-average power ratio (PAPR) [17, 20, 26, 27].

In SIM, the transmitter block diagrams for OFDM and *m*CAP are illustrated in figure 5.8. The modulation formats are identical to those illustrated in chapter 4 aside from the subcarrier index selector blocks and their receivers operate in the same manner, after the blocks have been split at the receiver. The data is split into $N_g = N/N_s$ groups of subcarriers, where N is the total number of subcarriers and N_s is the number of subcarriers per group. According to the desired light level, N_a subcarriers are activated and transmit information. The incoming bits are also grouped into N_g equivalent streams. In each of the groups $\left\lfloor \log_2 \binom{N_s}{N_a} \right\rfloor$ bits modulate

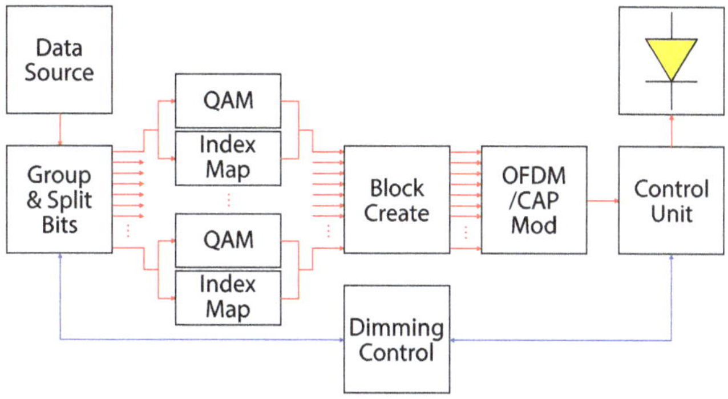

Figure 5.8. A generic transmitter block diagram for a subcarrier index modulated system. This block diagram includes the index map blocks that control which subcarriers are active and is generic for any multi-carrier system. Dimming control is provided by feedback through a control unit at the transmitter.

the N_a active subcarriers. It should be noted that $\binom{N_s}{N_a} = {}_{N_s}C_{N_a} = N_s![N_a!(N_s - N_a)!]^{-1}$, i.e. the number of combinations of N_s and N_a. The remaining bits modulate the QAM constellation used for the individual subcarriers. Therefore it is possible to state for the ith group [17]:

$$\Omega_{G_i} = \left\{\omega_1^{G_i}, \omega_2^{G_i}, \ldots, \omega_{N_a}^{G_i}\right\} \tag{5.11}$$

where Ω_{G_i} is the set of selected subcarrier indices for the ith group G_i, and $i \in \{1, 2, \cdots, G\}$. The selected subcarrier indices are given by $\omega_{n_a}^{G_i} \in \{1, 2, \cdots N\}$ and $n_a = \{1, 2, \cdots, N_a\}$. Therefore, for any two given sets of selected subcarriers, $\Omega_k \cap \Omega_j = \varnothing$.

The modulation of the subcarriers then happens as normal for (a) OFDM, insertion of the necessary Hermitian symmetry and the IFFT, and (b) pulse shaping for *m*CAP. A control unit exists to feedback the dimming level required at the point of transmission and to make adjustments to the number of active subcarriers.

In the OFDM formulation of the dimming scheme, analysis was performed in [17] that provides insight into the channel capacity whilst dimming is enabled. They adopt the reverse polarity OFDM structure previously outlined in (5.9) to ensure there is continuous transmission over the entire symbol period.

The mutual information within each group of subcarriers can be described using the chain rule of mutual information $I(.)$ according to [17]:

$$I(\mathbf{x_s}, \mathbf{x_c};\; \mathbf{y}) = I(\mathbf{x_s};\; \mathbf{y}|\mathbf{x_c}) + I(\mathbf{x_c};\; \mathbf{y}) \tag{5.12}$$

where $\mathbf{x_s}$ and $\mathbf{x_c}$ are the group-wise QAM constellation-domain and index-domain symbols, respectively. The above is the sum of these two which gives the total mutual information and hence, treatment of both parts is therefore required.

The constellation-domain information is given as [17]:

$$I(\mathbf{x}_s;\ \mathbf{y}|\mathbf{x}_c) = \sum_{j=1}^{\binom{N_s}{N_a}} \mathrm{P}(\mathbf{x}_c = \Psi_j) I(\mathbf{x}_s;\ \mathbf{y}|\mathbf{x}_c = \Psi_j) \tag{5.13}$$

where P(.) is the probability, and Ψ_j is the jth index-domain symbol, $j \in \left\{1, 2, \cdots, \binom{N_s}{N_a}\right\}$. If one makes the fair assumption that the subcarriers are activated with equal probability, $\mathrm{P}(\mathbf{x}_c = \Psi_j) = \binom{N_s}{N_a}^{-1}$. Using this, (5.13) becomes:

$$I(\mathbf{x}_s;\ \mathbf{y}|\mathbf{x}_c) = \frac{N_a}{N_s} \sum_{i=1}^{N_s} \log_2\left[1 + \frac{H_i^2 P_t}{2\sigma^2 N_g N_a}\right] \tag{5.14}$$

where P_t is the transmitted power, H_i is the channel coefficient of the ith subcarrier, and σ^2 is the noise variance.

Equivalently, the upper-bound of the index-domain mutual information can be approximated following the derivation in [17] as:

$$I(\mathbf{x}_c;\ \mathbf{y}) \approx \log_2\binom{N_s}{N_a} \tag{5.15}$$

Since the mutual information can be calculated by substituting (5.14) and (5.15) into (5.12), it is possible to calculate the capacity of the SIM-based asymmetrically-clipped OFDM dimming system, since [17]:

$$C = \frac{B}{N} N_g I(\mathbf{x}_s, \mathbf{x}_c;\ \mathbf{y}) = \frac{B}{4N_s}(I(\mathbf{x}_s;\ \mathbf{y}|\mathbf{x}_c) + I(\mathbf{x}_c;\ \mathbf{y})) \tag{5.16}$$

where B is the bandwidth. Substitution of (5.14) and (5.15) into (5.16) yields the final form of the capacity [17]:

$$C \approx \frac{BN_a}{4N_s^2} \sum_{i=1}^{N_s} \log_2\left[1 + \frac{H_i^2 P_t}{2\sigma^2 N_g N_a}\right] + \frac{B}{4N_s} \log_2\binom{N_s}{N_a} \tag{5.17}$$

Clearly, there is a high dependence on the SNR, (in this case proportional to $H_i^2 P_t \sigma^{-2}$), and the spectral efficiency is directly proportional to this value. There is also a dependence on the ratio of N_g, N_s and N_a. In figure 5.9, the capacity estimate defined in (5.17) is shown for a range of SNRs.

5.3 Light balancing

Dimming has been shown to cause variations in the emitted optical characteristics of the LED such as emitted wavelength and temperature (chromaticity) [9, 28, 29]. Balancing the colour of light that is used is also an important factor that has been

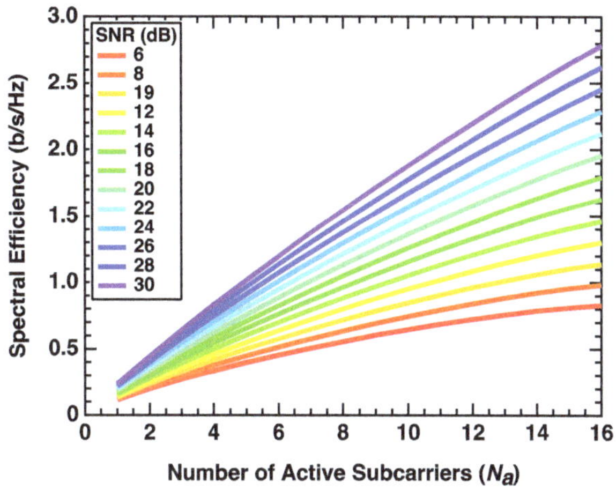

Figure 5.9. The spectral efficiency of the dimming system based on index modulated OFDM performed in [17], which shows that the spectral efficiency increases when more subcarriers become active.

investigated in VLC systems. The vast majority of investigations have been in the form of colour shift keying (CSK), however, there has been one key investigation into the impact of the quality of white light emission [9]. This publication is a milestone in VLC since it is the first step to forming a link between the quality of light emitted and the impact on human health.

The report argues that several of the conditions that are key to VLC operation such as LED temperature, drive current and wavelength can have an impact on a human's psychological and/or biological behaviours and is supported by the literature [9, 30, 31]. For instance, the production of the stress hormone, cortisol, can be promoted under certain lighting conditions. There have been numerous other studies that investigate the impact of indoor lighting on seasonal affective disorder [32], the impact of lighting on gender and age [33] and even the perception of personal space under differing light conditions [34].

Clearly, this is an area that VLC systems must consider as a topic of importance. In [9] it is reported that cortisol and melatonin, the hormones that control stress and sleep pattern are heavily influenced by blue light. This is a particularly important consideration because a considerable quantity of literature reports the use of InGaN (blue-emitting) colour-converted LEDs. The European Union has also investigated the use of artificial light and its impact on health [35], and linked disruptions of the above hormones to early onset of chronic illness such as several types of cancer, among others. In general, LED manufacturers set limits to their drive conditions of their devices that have considered such health implications, and when driven under constant conditions, can minimise the risk. However, in VLC systems, information transmission occurs via IM, which effectively means deviating the current across the diode from the dc bias continuously, which in turn results in small deviations of the emitted light intensity and chromaticity, which has linkage to the above. The report in [9] examines the effect in the shift in these quantities and their impact on the 'white'-light emitted from the LED. It is reported in [9] that a shift in chromaticity occurs when signals are added to the dc bias driving the LED and hence, modulation formats must be pre-balanced in order to avoid this effect. One of the key modulation formats that operates on the chromaticity of the emitted light is CSK.

5.3.1 Colour shift keying

In the early IEEE 802.15.7 VLC physical layer standard, CSK was adopted, originally making use of an RGB-LED [36] as the source, and more recent reports have extended to RAGB-LEDs [37–40]. In the original system, significant work was done to investigate the performance of CSK. The basic idea is that information is embedded into the chromaticity of the LEDs while the overall light intensity remains constant to avoid flickering and consider health requirements such as eye safety. In contrast to the usual approach to modulation, the symbols are encoded in the RGB space, as opposed to the signal space.

The block diagram for the CSK system is illustrated in figure 5.10. The first step is to convert the incoming bits into $(x, \ y)$ coordinates that map to the Commission Internationale de l'Éclairage (CIE) 1931 xyY colourspace [41], which is illustrated in figure 5.11. The coordinates correspond to a location on that colourspace, which represents a certain chromaticity. For a three-colour LED, a set of coordinates are generated before conversion to an RGB intensity.

The CSK alphabet is given by $\Lambda = \{\mathbf{P}_0, \ \mathbf{P}_1, \ \cdots, \ \mathbf{P}_{M-1}\}$ and the ith symbol is $\mathbf{P}_i = \left(P_r, \ P_g, \ P_b\right)$, the components of which represent the power transmitted by the

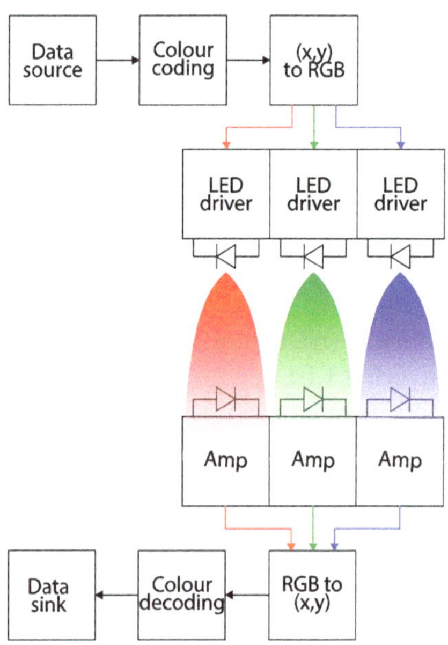

Figure 5.10. Generic block diagram of a CSK system. A new block that converts traditional (x,y) coordinates into RGB values is introduced at the transmitter and receiver to convert the information into the colour domain.

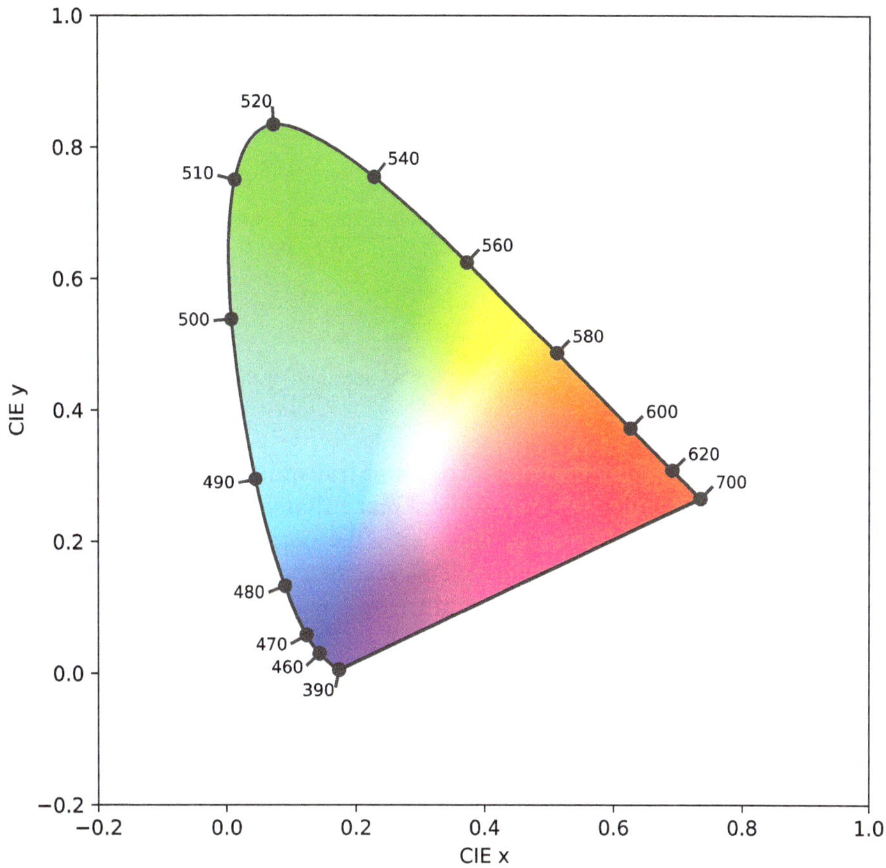

Figure 5.11. The CIE 1931 xyY colourspace with visible wavelengths highlighted around the edge of the colour gamut.

individual RGB-LEDs. As discussed in (3.15) in chapter 3, the received signal is developed to [41]:

$$\mathbf{r} = \mathfrak{R}\mathbf{H}(0)\mathbf{P} + \mathbf{n} \tag{5.18}$$

where $\mathbf{H}(0) = \left(H_r(0),\quad H_g(0),\quad H_b(0)\right)$ is a vector that represents the channel gain for each colour and $\mathbf{n} = (n_r,\quad n_g,\quad n_b)$ is the AQGN for each colour. The responsivity $\mathfrak{R}$ must also be considered, because it is not constant for each wavelength when a silicon diode is used with RGB wavelengths (refer to chapter 2). The individual components of any alphabet element $\mathbf{P}$ have following constraint [41]:

$$P_r \geqslant 0, \quad P_g \geqslant 0, \quad P_b \geqslant 0 \tag{5.19}$$

and the total intensity $P_t = P_r + P_g + P_b$.

Before the demodulation is discussed, it is first necessary to provide a light treatment to the three-dimensional signal space and its conversion to a two-dimensional projection. The three-dimensional signal space is illustrated in

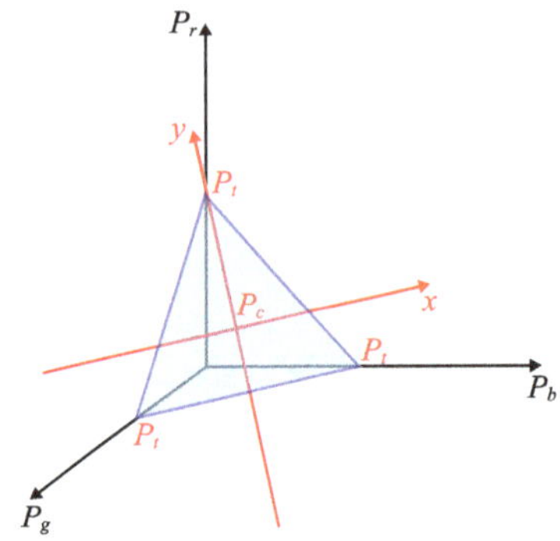

Figure 5.12. The intersection of the possible CSK symbol locations within its symbol space with the conventional Cartesian plane.

figure 5.12. Clearly, retaining that the total power transmitted is the sum of the three components, there is a finite two-dimensional plane that can be defined in the three-dimensional space. The extremes of the plane, the corners of the two-dimension triangle, are given when only a single element is active and the rest are off. The centroid of the two-dimensional equilateral triangle P_c is used as the reference point and the line passing through the vertex $(0, \ P_t, \ 0)$ is set as the y-axis. Using this information alongside simple geometry and trigonometry, to calculate the centroid position as $P_c = (P_t/3, \ P_t/3, \ P_t/3)$, while the x-axis is given as $P_r + P_b = 2P_t/3$, $P_g = P_t/3$. Following which, a transformation matrix **T** can be generated to translate between the three-dimensional signal space and the more traditional Cartesian plane [42, 43]:

$$\mathbf{T} = \begin{bmatrix} \frac{1}{\sqrt{2}} & 0 & -\frac{1}{\sqrt{2}} \\ -\frac{1}{\sqrt{6}} & \frac{2}{\sqrt{6}} & -\frac{1}{\sqrt{6}} \end{bmatrix} \tag{5.20}$$

It is relatively intuitive that the signal space increases with the transmission power, enlarging the equilateral triangle the contains the possible symbols. The symbols may therefore be represented in two-dimensional Cartesian space as follows [41]:

$$\bar{\mathbf{P}} = \mathbf{TP} \tag{5.21}$$

This concept is illustrated in figure 5.13, and it should be noted that now working in the Cartesian space maintains compatibility with the more CIE 1931 xyY colourspace shown above in figure 5.11. At the receiver, the reverse process occurs, first starting with the traditional minimum Euclidean distance method based on the ith received symbol $\mathbf{r}_i$ [41, 43]:

$$\bar{\mathbf{r}}_i = \underset{\mathbf{r}_i \in \Lambda}{\arg\min} \| \mathbf{r}_i - \mathbf{P} \| \tag{5.22}$$

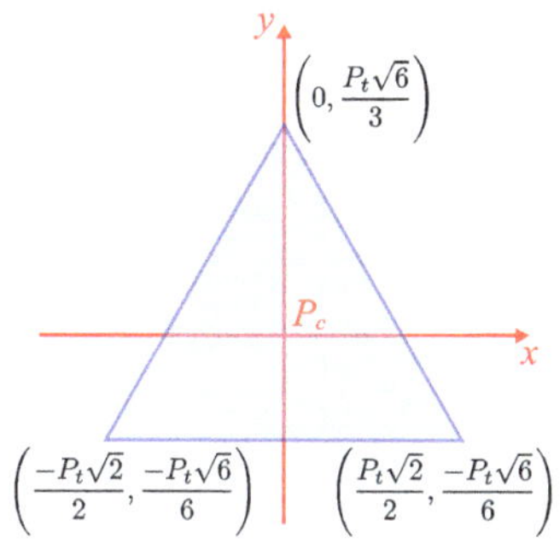

Figure 5.13. Respective values of the extremes of the CSK symbol space.

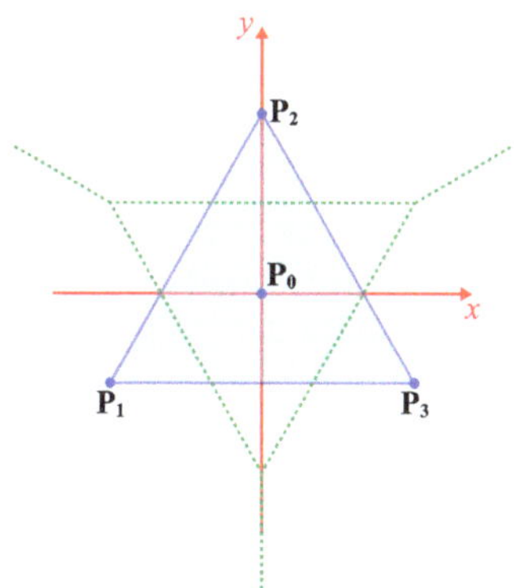

Figure 5.14. The dashed green line shows an example of possible threshold values within the symbol space, while other thresholds could be used, this one provides the maximum Euclidean distance for the 4-CSK system shown within the figure.

It is also possible to form decision boundaries based on the minimum Euclidean distance, which are illustrated in figure 5.14. It is noteworthy that there is also a datapoint positioned at $\mathbf{P}_c$. Once the symbols are estimated, the next step of the demodulation process is to de-map the RGB symbols to the xy coordinates and then de-map the bits back to the binary stream. In order to calculate the BER performance of CSK, it is necessary to determine the probability of transition from the transmitted symbol s_t to the received one s_r. Based on derivations found through the combination of [41] and [44], the following is derived [41, 43]:

$$\mathrm{P}(s_r|s_t) = \iint_{\mathcal{D}_k} s_{\mathrm{pdf}}(m, \theta)\mathrm{d}m\mathrm{d}\theta \tag{5.23}$$

where P(.) is the probability, $\mathcal{D}_k$ is the decision region of interest, and s_{pdf} is the probability density function (PDF) of the noise, which is expected to be dominated by AWGN and is given with magnitude m and phase θ by [41, 43]:

$$s_{\mathrm{pdf}}(m, \theta) = \frac{m}{\pi N_0}\exp\left(-\frac{m^2}{N_0}\right) \tag{5.24}$$

The relation between the received symbol and the decision region of interest is a an integral relating to (5.24). In [43], the detailed derivation is performed and can be observed, however, in this overview, we skip to the probability of receiving the symbol S_l, if the decision region is $\mathcal{D}_1$ as follows [43]:

$$P(\mathcal{D}_1|S_l) = \int_{\theta_1}^{\theta_2} \int_{R(\theta)}^{\infty} s_{\mathrm{pdf}}(m, \theta)\ \mathrm{d}r\mathrm{d}\theta \tag{5.25}$$

where $R(\theta)$ is the distance from the symbol S_l to the centre of the decision region boundary.

The exact BER can be given by the summation of all the possible transitions and their probabilities of error as follows [43, 44]:

$$\mathrm{BER}_{\mathrm{AWGN}} = \frac{1}{\log_2 M} \sum_{l=0}^{M-1} \sum_{\substack{k=0 \\ k\neq l}}^{M-1} \mathrm{d}(S_l, S_k) P(S_k|S_l) P(S_l) \tag{5.26}$$

where $\mathrm{d}(S_l, S_k)$ is the Hamming distance between the two symbols [42]. The BER performance of an M-CSK system is illustrated in figure 5.15 for $M = 4$, 8 and 16. Similar BER profiles can be observed in comparison to traditional modulation formats such as PAM.

5.3.2 Coloured pulse amplitude modulation

Adding multiple-users is an important consideration and the opportunity is enhanced when using multiple wavelengths. In general, WDM is the standard mode to improve data rates by transmitting different information on each wavelength, where the wavelengths are spaced far enough part to avoid crosstalk. Normally, the number of receivers must match the number of wavelengths used and dichroic filters are used to isolate the selected wavelength at the receiver,

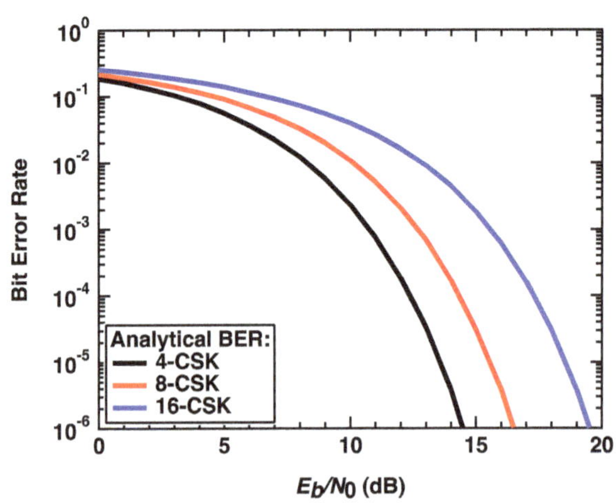

Figure 5.15. The BER performance of a range of CSK systems, showing that a BER of 10^{-6} can be obtained for E_b/N_0 values of ~14.5, 16.5 and 19.5 dB, for 4-, 8- and 16-CSK, respectively.

ensuring only the desired colour is recovered for demodulation. This approach has been well documented in the literature [45–48] and has generally been used to achieve the highest data rates. In addition, CSK has similar requirements in terms of the number of receivers and dichroic filters. High quality dichroic filters with sharp cut-on and cut-off frequencies are expensive and must be designed for the RAGB wavelengths used for transmission.

In [49], a new approach was taken that removes most of the complexity of the system and ensures that multiple wavelengths can be recovered simultaneously, using the naturally broadband nature of the silicon absorption region. The new scheme, termed coloured PAM (CPAM) allows the independent symbols from multiple wavelengths to superpose in free-space before absorption by a single detector. The symbols can be identified via hard threshold detection at the receiver via a classification table as will be described, by intentionally weighting the signal amplitudes according to wavelength.

The system block diagram is illustrated in figure 5.16. In the general form, each LED is denoted from LED_0 to LED $_{n-1}$ where n is the number of wavelengths present in the system, in order of the combination of their total optical power output and responsivity. Each LED has a relative amplitude weight of 2^0 to 2^{n-1} in line with their respective indices, and therefore it is clear that the relative signal amplitudes are doubled with each wavelength added to the system. The total number of levels in the system is $M = 2^n$. In the report presented in [49], the wavelengths of the RGB-LED used were 469 nm (blue), 529 nm (green) and 645 nm (red) and those are highlighted in the silicon responsivity curve in figure 5.17(a) and their responsivities are measured as ~0.26 A/W, 0.32 A/W and 0.45 A/W, respectively. If all the LED chips are transmitted with the same amplitude, logic dictates that they would have three individual photocurrent amplitudes after reception, but they are not evenly spaced for the LED used.

Therefore, the electro-optic response is also factored in (see figure 5.17(b)) as the optical power emission of the LED chips are also uneven. The relative amplitude weights are defined according to the combined response of these two characteristics and the modulation depth. The signal with the highest modulation depth is the one that is assigned the highest wavelength.

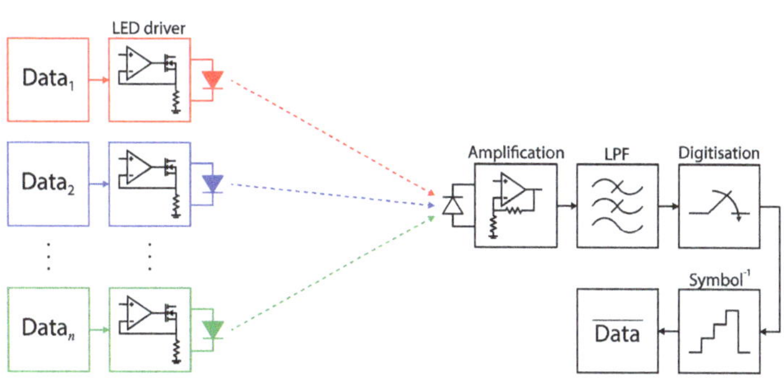

Figure 5.16. The block diagram for the CPAM system proposed in [49], where only one PD is present with no colour filters.

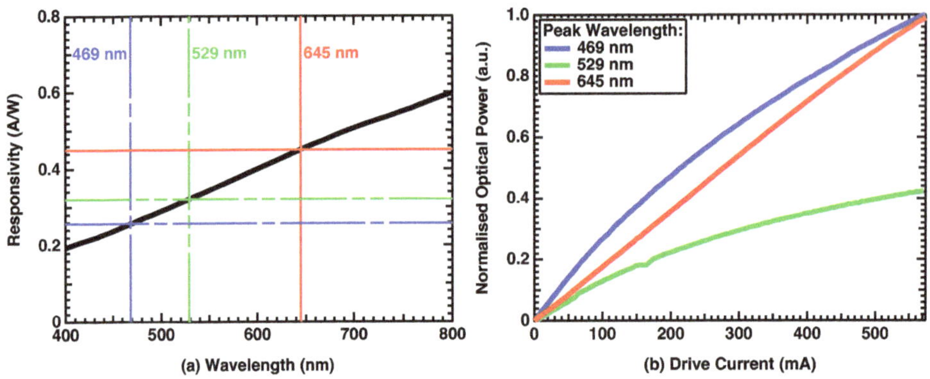

Figure 5.17. (a) The PD responsivity with the peak wavelengths of the LEDs used highlighted and (b) the same LED current-optical power relationships.

For instance, in the case presented in [49], the weightings are assigned 2^2 to red, since it has the highest combination of optical power and responsivity, 2^1 to blue, since it has the second highest and finally, 2^0 to the green because it has the lowest combination. The order of the weights are found mathematically after measuring the wavelengths and electro-optic responses of the LEDs.

Next, the weights are applied to the data as follows [49]:

$$y_R(t) = [x_R(t)w_R \otimes h(t)]\Re(645) + n(t) \tag{5.27}$$

where x_R and w_R are the data on and weighting of the red wavelength, respectively, $h(t)$ is the channel response, $\Re(\lambda)$ is the wavelength and $n(t)$ is AWGN. Likewise, for the green and blue wavelengths [49]:

$$y_G(t) = [x_G(t)w_G \otimes h(t)]\Re(529) + n(t) \tag{5.28}$$

$$y_B(t) = [x_B(t)w_B \otimes h(t)]\Re(469) + n(t) \tag{5.29}$$

The total received signal in general form is [49]:

$$y_{\text{CPAM}}(t) = \sum_{\lambda=0}^{n-1} y_\lambda(t) \tag{5.30}$$

Demodulation of the CPAM signal is done in two straightforward steps according to the M signal levels observed and operates in the same manner as conventional PAM. The M number of CPAM symbol levels correspond to individual bit levels as described in tables 5.2 and 5.3, for two- and three-wavelength systems, respectively.

The LEDs used in [49] were equalised using simple cascaded resistor-capacitor equalisers and all have bandwidths approximately around 10 MHz. For a system employing a red and blue LED, individual data rates of approximately 10 MBd were recovered, indicating a data rate of 20 Mb/s using a single receiver at a BER of 10^{-6}, as illustrated in figure 5.18(a).

Table 5.2. The symbol map for 2-wavelength CPAM.

Symbol	Red	Blue
0	0	0
1	0	1
2	1	0
3	1	1

Table 5.3. The symbol map for 3-wavelength CPAM.

Symbol	Red	Blue	Green
0	0	0	0
1	0	0	1
2	0	1	0
3	0	1	1
4	1	0	0
5	1	0	1
6	1	1	0
7	1	1	1

When the green LED is added, the individual BER performance drops to ~5 MBd per wavelength at a similar BER as previously, which is a slight improvement in aggregated data rate, although the individual baud rates have dropped (see figure 5.18(b)). This is due to the additional SNR requirements for each link and the reduction in Euclidean distance. If the electro-optic responses of the LEDs were higher power with sharper gradient and/or likewise the PD responsivity, instinct implies that the improvement would be greater.

5.4 Summary

This chapter has looked at balancing lighting requirements with data communications. As VLC is constantly discussed as a dual purpose, dual advantage technology, effort must be made to provide high quality lighting systems otherwise there will never be market penetration, if the illumination capacity is insufficient. The relationships between measured and perceived light are non-linear and have been discussed at the start of the chapter. This is particularly important when designing an illumination system because control of the light level must adhere to several standards.

A generic block diagram was given for a dimming-enabled communication system and discussed along with the required illumination levels for several scenarios. Dimming capabilities were previously enabled by several different PWM approaches, which were discussed in this chapter, including rate- and level-

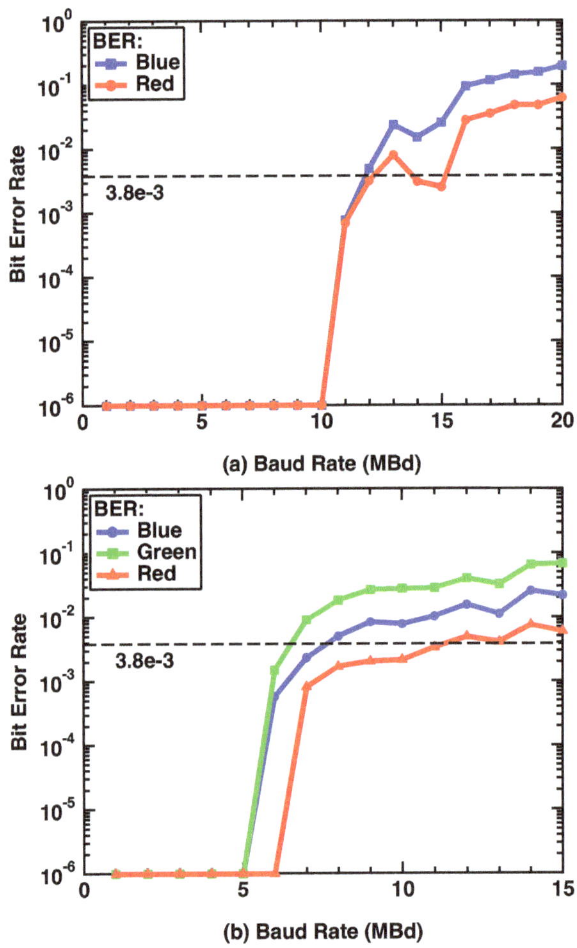

Figure 5.18. The BER performance for (a) 2-wavelength CPAM and (b) 3-wavelength CPAM. A higher aggregate data rate can be obtained using the latter system; however, due to decreased Euclidean distance, lower individual data rates can be obtained, problematically.

adaptive systems. The disadvantage of most PWM approaches to dimming was that a punctured data rate must generally be introduced.

Eventually, advanced systems were proposed by Elgala *et al*, who introduced reverse polarity OFDM that meant that the data could always be transferred by flipping the polarity of the signal when the PWM signal is 'off'. This was a major leap forward for dimming systems. Other methods discussed include PPM and SIM.

The second half of the chapter looks at colour-based modulation, in particular CSK and coloured PAM. The more popular of the two is CSK where a large amount of theoretical work has been proposed to date, along with a number of high quality experimental demonstrations. The general approaches to translating conventional data to the CSK domain, and how to make symbol decisions at the receiver, are outlined in this chapter.

References

[1] Rea M S *et al* 2000 *The IESNA Lighting Handbook: Reference and Application*

[2] Zafar F, Karunatilaka D and Parthiban R 2015 Dimming schemes for visible light communication: the state of research *IEEE Wireless Commun.* **22** 29–35

[3] Dowhuszko A A, Ilter M, Pinho P and Hämäläinen J 2020 The effect of power allocation on visible light communication using commercial phosphor-converted LED lamp for indirect illumination *ICASSP 2020 IEEE Int. Conf. on Acoustics, Speech and Signal Processing (ICASSP)* (IEEE) pp 5225–9

[4] Komine T and Nakagawa M 2004 Fundamental analysis for visible-light communication system using LED lights *IEEE Trans. Consum. Electron.* **50** 100–7

[5] Schubert E F 2006 Human eye sensitivity and photometric quantities *Light-Emitting Diodes* (Cambridge: Cambridge University Press) pp 275–91

[6] Lee K and Park H 2011 Modulations for visible light communications with dimming control *IEEE Photonics Technol. Lett.* **23** 1136–8

[7] Shannon C E 1948 A mathematical theory of communication *Bell Syst. Tech. J.* **27** 379–423

[8] Deng P, Kavehrad M and Kashani M A 2015 Nonlinear modulation characteristics of white LEDs in visible light communications *Optical Fiber Communication Conf.* (Optical Society of America) p W2A-64

[9] Popoola W O 2016 Impact of VLC on light emission quality of white LEDs *J. Lightwave Technol.* **34** 2526–32

[10] Choi J- H, Cho E- B, Ghassemlooy Z, Kim S and Lee C G 2015 Visible light communications employing PPM and PWM formats for simultaneous data transmission and dimming *Opt. Quantum Electron.* **47** 561–74

[11] Jang H-J, Choi J-H, Ghassemlooy Z and Lee C G 2012 PWM-based PPM format for dimming control in visible light communication system *2012 8th Int. Symp. on Communication Systems, Networks & Digital Signal Processing (CSNDSP)* (IEEE) pp 1–5

[12] Jang H-J, Choi J-H and Lee C G 2011 Simulation of a VLC system with 1 Mb/s NRZ-OOK data with dimming signal *IET Conf. Proc.*

[13] Kizilirmak R C and Kho Y H 2015 Mitigation of illumination interference caused by PWM dimming in OFDM based visible light communication systems *2015 Int. Conf. on Computer, Communications, and Control Technology (I4CT)* (IEEE) pp 489–92

[14] Bai B, Xu Z and Fan Y 2010 Joint LED dimming and high capacity visible light communication by overlapping PPM *19th Annual Wireless and Optical Communications Conf. (WOCC 2010)* (IEEE) pp 1–5

[15] Rajagopal S, Roberts R D and Lim S-K 2012 IEEE 802.15.7 visible light communication: modulation schemes and dimming support *IEEE Commun. Mag.* **50** 72–82

[16] You X, Chen J, Zheng H and Yu C 2015 Efficient data transmission using MPPM dimming control in indoor visible light communication *IEEE Photonics J.* **7** 1–12

[17] Wang T, Yang F, Pan C, Cheng L and Song J 2019 Spectral-efficient hybrid dimming scheme for indoor visible light communication: A subcarrier index modulation based approach *J. Lightwave Technol.* **37** 5756–65

[18] Wang C, Yang Y, Cheng J, Guo C and Feng C 2019 A dimmable OFDM scheme with dynamic subcarrier activation for VLC *IEEE Photonics J.* **12** 1–12

[19] Akande K O and Popoola W O 2019 Enhanced subband index carrierless amplitude and phase modulation in visible light communications *J. Lightwave Technol.* **37** 5867–74

[20] Akande K O and Popoola W O 2018 Subband index carrierless amplitude and phase modulation for optical communications *J. Lightwave Technol.* **36** 4190–7

[21] Proakis J G and Salehi M 2007 *Fundamentals of Communication Systems* (New Delhi: Pearson Education India)

[22] Elgala H and Little T D C 2013 Reverse polarity optical-OFDM (RPO-OFDM): dimming compatible OFDM for gigabit VLC links *Opt. Express* **21** 24288–99

[23] Cho E, Choi J-H, Park C, Kang M, Shin S, Ghassemlooy Z and Lee C G 2011 NRZ-OOK signaling with LED dimming for visible light communication link *2011 16th European Conf. on Networks and Optical Communications* (IEEE) pp 32–5

[24] Armstrong J and Schmidt B J C 2008 Comparison of asymmetrically clipped optical OFDM and dc-biased optical OFDM in AWGN *IEEE Commun. Lett.* **12** 343–5

[25] Armstrong J and Lowery A J 2006 Power efficient optical OFDM *Electron. Lett.* **42** 370–2

[26] Islim M S, Tsonev D and Haas H 2015 Spectrally enhanced PAM-DMT for IM/DD optical wireless communications *2015 IEEE 26th Annual Int. Symp. on Personal, Indoor, & Mobile Radio Communications (PIMRC)* (IEEE) pp 877–82

[27] Islim M S and Haas H 2016 Augmenting the spectral efficiency of enhanced PAM-DMT-based optical wireless communications *Opt. Express* **24** 11932–49

[28] Manninen P and Orreveteläinen P 2007 On spectral and thermal behaviors of AlGaInP light-emitting diodes under pulse-width modulation *Appl. Phys. Lett.* **91** 181121

[29] Gu Y, Narendran N, Dong T and Wu H 2006 Spectral and luminous efficacy change of high-power LEDs under different dimming methods *Proc. SPIE* **6337** 63370J

[30] Riemersma-Van Der Lek R F, Swaab D F, Twisk J, Hol E M, Hoogendijk W J G and Van Someren E J W 2008 Effect of bright light and melatonin on cognitive and noncognitive function in elderly residents of group care facilities: a randomized controlled trial *JAMA* **299** 2642–55

[31] Hankins M W, Peirson S N and Foster R G 2008 Melanopsin: an exciting photopigment *Trends Neurosci.* **31** 27–36

[32] Tonello G 2008 Seasonal affective disorder: Lighting research and environmental psychology *Light. Res. Technol.* **40** 103–10

[33] Knez I and Kers C 2000 Effects of indoor lighting, gender, and age on mood and cognitive performance *Environ. Behav.* **32** 817–31

[34] Adams L and Zuckerman D 1991 The effect of lighting conditions on personal space requirements *J. Gen. Psychol.* **118** 335–40

[35] Mattsson M-O, Jung T and Proykova A 2011 Health effects of artificial light *Scientific Committee on Emerging and Newly Identified Health Risks (SCENIHR), Technical Report*

[36] Monteiro E and Hranilovic S 2014 Design and implementation of color-shift keying for visible light communications *J. Lightwave Technol.* **32** 2053–60

[37] Zuo Y, Zhang J and Qu J 2019 Power allocation optimization design for the quadrichromatic LED based VLC systems with illumination control *Crystals* **9** 169

[38] Dong J-M, Zhu Y-J and Sun Z-G 2019 Adaptive multi-color shift keying constellation design for visible light communications considering lighting requirement *Opt. Commun.* **430** 293–8

[39] Xiao Y and Zhu Y-J 2019 Chromaticity-adaptive generalized spatial modulation for MIMO VLC with multi-color LEDs *IEEE Photonics J.* **11** 1–12

[40] Xiao Y, Zhu Y-J, Zhang Y-Y and Sun Z-G 2018 Linear optimal signal designs for multi-color MISO-VLC systems adapted to CCT requirement *IEEE Access* **6** 75519–30

[41] Jia L, Wang J-Y, Zhang W, Chen M and Wang J-B 2015 Symbol error rate analysis for colour-shift keying modulation in visible light communication system with RGB light-emitting diodes *IET Optoelectron.* **9** 199–206

[42] Tzikas A E and Sahinis A 2019 3-color shift keying for indoor visible light communications *IEEE Commun. Lett.* **23** 2271–4

[43] Tang J, Zhang L and Wu Z 2018 Exact bit error rate analysis for color shift keying modulation *IEEE Commun. Lett.* **22** 284–7

[44] Xiao L A new approach to calculating the exact transition probability and bit error probability of arbitrary two-dimensional signaling *IEEE Global Telecommunications Conf. 2004 GLOBECOM '04* vol 2 pp 1239–43

[45] Bian R and Tavakkolnia I 2019 15 15.73 Gb/s visible light communication with off-the-shelf LEDs *J. Light. Technol.* **37** 2418–24

[46] Wang Y, Tao L, Huang X and Shi J 2015 8-Gb/s RGBY LED-based WDM VLC system employing high-order CAP modulation and hybrid post equalizer *IEEE Photonics J.* **7** 1–7

[47] Zhu X, Wang F, Shi M, Chi N and Liu J 2018 10.72Gb/s visible light communication system based on single packaged RGBYC LED utilizing QAM-DMT modulation with hardware pre-equalization *2018 Optical Fiber Communications Conf. and Exposition (OFC)* pp 1–3

[48] Chun H, Gomez A, Quintana C, Zhang W, Faulkner G and O'Brien D 2019 A wide-area coverage 35Gb/s visible light communications link for indoor wireless applications *Sci. Rep.* **9** 4952

[49] Burton A, Haigh P A, Chvojka P, Ghassemlooy Z and Zvanovec S 2019 Filter-less WDM for visible light communications using colored pulse amplitude modulation *Opt. Lett.* **44** 4849–52

Chapter 6

Uplink technologies

6.1 Introduction

The natural layout of rooms with lighting fixtures on the ceilings leads to a major open challenge in VLC research. As lights are usually positioned on the ceiling facing downwards, the data uplink does not have a natural solution, since in general people are reluctant to have upward facing lamps on the floor that they could connect with. Therefore, researchers have experimented with multi-technology networks that include a different technology as the uplink.

As well as visible light, researchers have proposed IR, Wi-Fi and RF as alternative uplink solutions. Such technologies can be proposed because they are invisible and typically the uplink capacity is around 10% of the downlink in 5G [1]. In this chapter, the state-of-the-art uplink proposals are reviewed.

6.1.1 Visible light uplink

In spite of the fact that it is relatively unusual to have upward facing lamps, it is also possible to utilise upward facing display technologies among other options and hence research has been performed into VLC uplinks. The first demonstration of a fully integrated bi-directional VLC network was by the author in [2], which demonstrated a 10BASE-T Ethernet network. The packets are transmitted using the user datagram protocol (UDP) [3], which is generally considered as unreliable because there is no acknowledgement from the receiver that a packet has been successfully received, unlike transmission control protocol (TCP) where packets have sequence numbers that must be acknowledged before further transmission occurs [4].

The quality of service (QoS) was measured in [2] to understand the performance of the link. The transceiver bandwidth was limited to ~8 MHz using a blue filter, which is sufficient to support a data rate of 10 Mb/s, free of ISI. The QoS was interpreted using key performance indicators such as throughput, packet loss, BER, latency and jitter. In the 10BASE-T Ethernet standard [5], the maximum data

doi:10.1088/978-0-7503-1680-4ch6

payload size is 1500 bytes, and therefore packet sizes ranging from 1 byte to 1500 bytes were tested.

The BER and illuminance is illustrated as a function of distance in figure 6.1, as is the total jitter in figure 6.2. The BER increases with distance as expected and the illuminance drops, roughly with the expected inverse square of the distance. The BER never increases beyond a level that would cause link outage, however, the illuminance drops beneath the required level when the distance increases beyond approximately 0.5 m, indicating the limitations of the work.

Finally, the throughput is shown in figure 6.3, and it is clear that no matter the packet size, the throughput is steady at around 9 Mb/s, close to the maximum offered by the 10BASE-T standard. Interestingly, at around 50 bytes/packet, there is an 1 Mb/s drop in throughput. The report in [2] attributes this to a resonant frequency causing a mismatch in power transfer at the packet size, since small packet transmission is typically rare.

To complete this subsection, an example of a complete report from one of the world's leading VLC research groups at *Fudan University* is analysed. Integrated fibre-wireless downlinks have been presented in the literature extensively by either splitting the fibre [6, 7], RF [8–10] or VLC [11, 12] as the wireless network integration, as well as demonstrations of FSO-VLC [13–15]. However, in the vast majority of cases these are unidirectional links that operate in broadcast mode.

In [10], a high-speed VLC network was demonstrated, providing a total of 8 Gb/s to eight users in a star topology. The network consists of eight access points in full-duplex mode, with the uplink and downlink both supporting 500 Mb/s data rates. The full network architecture is illustrated in figure 6.4 and includes a fibre connection from a central office and the transmission utilises time-division multiple

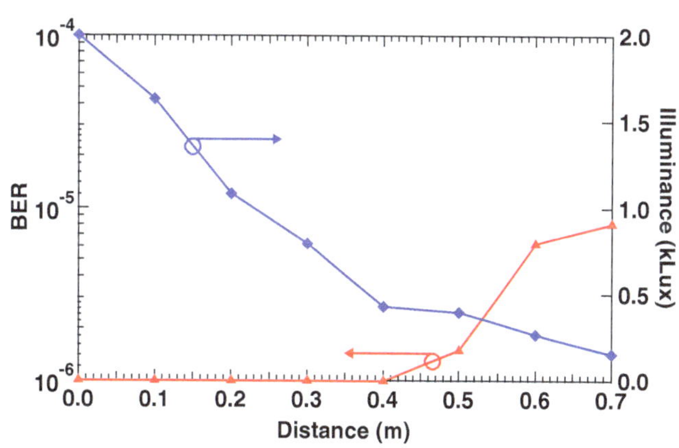

Figure 6.1. Measured illumination and BER as a function of distance for the system presented in [2]. © 2020 IEEE. Reprinted, with permission, from [2].

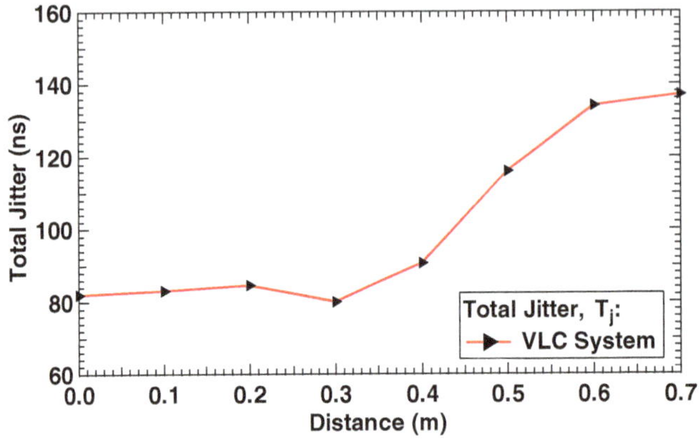

Figure 6.2. Measured total jitter as a function of distance for the system presented in [2]. © 2020 IEEE. Reprinted, with permission, from [2].

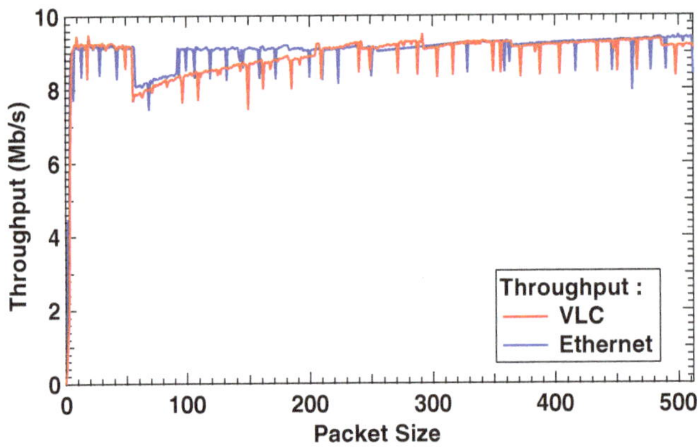

Figure 6.3. Measured throughput for the VLC system presented in [2] in comparison to wired Ethernet as a function of packet size. © 2020 IEEE. Reprinted, with permission, from [2].

access (TDMA) and frequency bands to make effective use of the wide optical fibre bandwidth. Each VLC access point within the house is allocated a frequency band on the fibre, and the time divisions are allocated to the uplink and downlink, respectively.

The down and uplink wavelengths used in the fibre are 1 548.49 nm and 1 545.04 nm, respectively, and they are intensity modulated by the summed data from all users in both directions. The modulation format used to transfer the information across the fibre is OFDM, although the particular details of the configuration are not revealed. The centre frequencies of the bands allocated to each of the eight users are 62.5 MHz, 87.5 MHz, 312.5 MHz, 437.5 MHz, 1062.5 MHz, 1565.5 MHz, 2062.5 MHz and 2437.5 MHz. Each frequency band is 100 MHz in bandwidth and a small 12.5 MHz buffer is used at the low frequencies. The length of the fibre

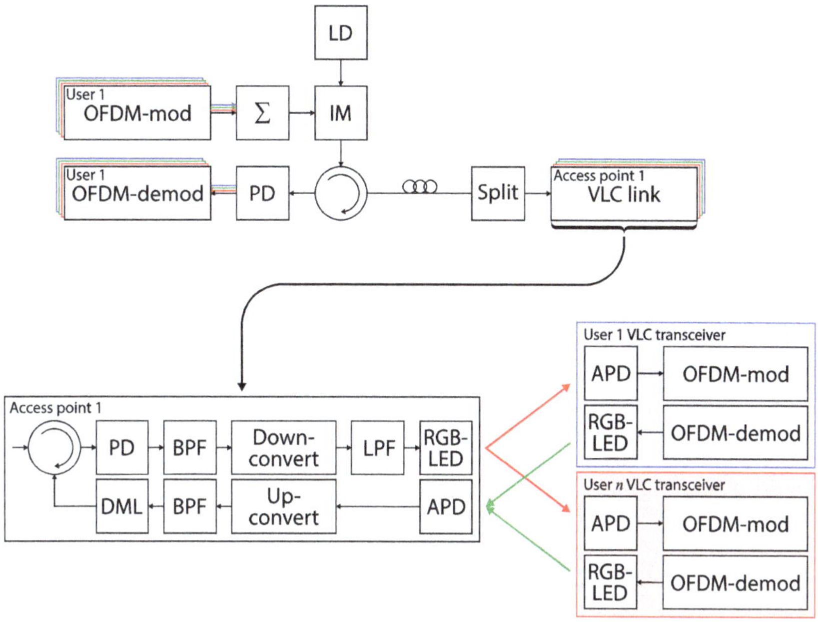

Figure 6.4. The complete architecture of the fibre-integrated VLC network presented in [10]. The fibre backbone connects the backbone via OFDM and each VLC access point connects to a number of users via modulation of a red LED in the downlink and a green one in the uplink.

transmission is 25 km over standard single-mode fibre. The modulation used is 32-QAM to ensure that the 100 MHz bandwidth carries a data rate of 500 Mb/s, which is the VLC up/downlink speed, as mentioned.

To make the signals compatible with the VLC access points, the signals from the fibre must be first detected using a PD and down-converted from the carrier frequency to baseband. This is done using a BPF and a local oscillator, before low-pass filtering. The down-converted signal is then biased and modulates the red (620 nm) wavelength of an RGB-LED in each access point. At the receiver, which is spaced 0.65 m from the transmitted, an avalanche PD (APD) collects the red-filtered light for demodulation. Standard OFDM demodulation is applied. Focusing lenses are used at both the transmitter and receiver to concentrate the optical power towards the detector of interest. The VLC uplink consists of the same circuit as the transmitted, however, this time the green (520 nm) chip of the RGB-LED is modulated.

The BER results of the link are illustrated in figure 6.5 for all of the different users and clearly similar error performance is obtained for all of them, as would be expected. There is a generally positive trend in the BER because the centre frequency of the bands assigned to the users increases with the user index and there is a slight frequency-dependent attenuation in the components the authors used to transfer the information.

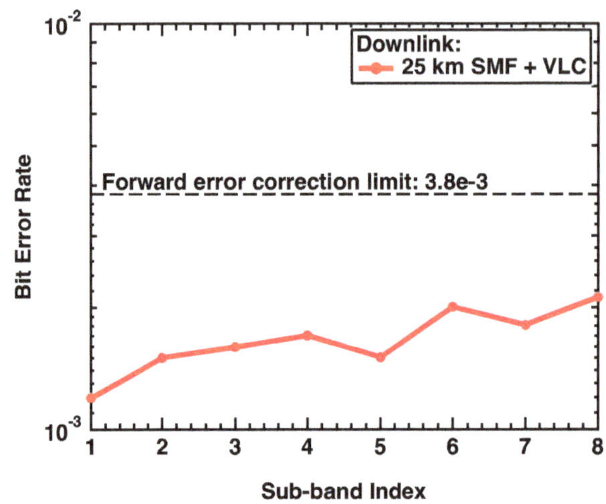

Figure 6.5. The measured BER of the fibre and VLC downlink of the integrated network demonstrated in [10].

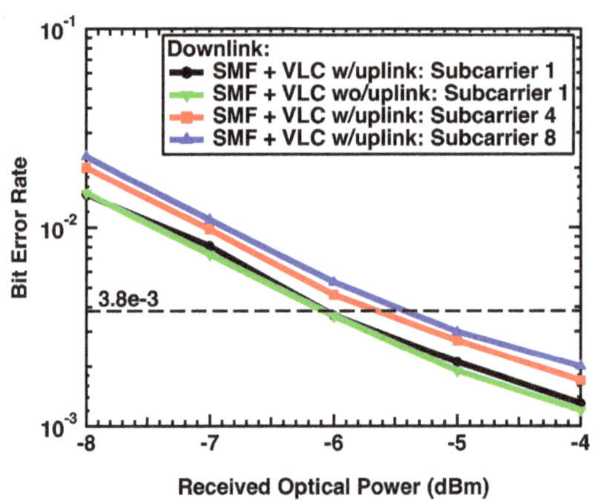

Figure 6.6. The measured individual subcarrier BERs for the fibre and VLC downlink as a function of received optical power of the integrated network demonstrated in [10].

In figure 6.6, the BER is illustrated as a function of the received optical power from the fibre transmission. Clearly, there is a minimum threshold of optical power required in order to provide sufficient SNR to the VLC access points, at around −5 dBm. The uplink clearly has no impact on the link performance, as illustrated by the performance of subcarrier 1 without and with the uplink present in the measurement.

Finally, figure 6.7 shows the BER performance of the VLC link as a function of distance and a received optical power from the fibre of −4 dBm. The link can support 500 Mb/s data rate transfer up to a distance of ~0.8 m, where it exceeds the maximum allowed BER.

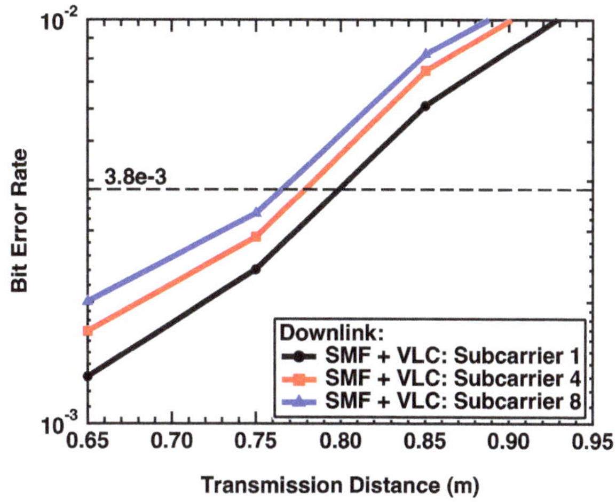

Figure 6.7. The measured individual subcarrier BERs for the fibre and VLC downlink as a function of free-space transmission distance of the integrated network demonstrated in [10].

6.1.2 Infrared uplink

One of the first examples of an IR uplink was reported in [16] in 2014, which also showed a fully working 10BASE-T Ethernet connection. Until this report, there was little focus on integrated systems, although high data rates exceeding 1.1 Gb/s had already been demonstrated in highly optimised, static links [17]. The difficulty in implementing real, mobile links at the time was mainly focused on the low bandwidths of the LEDs and fact that basic resistive-loading based equalisers were commonplace, which restricted optical power and therefore put strong limits on the achievable transmission distances (see chapter 3 and [18]).

The link was presented as an optical wireless bridge that connected two PCs instead of a wired or Wi-Fi connection. The VLC downlink was provided by 20 white LEDs divided into five groups of four devices in the configuration illustrated in figure 6.8. To ensure impedance matching between the first PC and the LEDs in conjunction with bandwidth extension, the transmitter consists of a buffer and passive equaliser. The functionality of which matches the 100 Ω PC output impedance with the 15 Ω LED input impedance. The equaliser deployed was a three-stage cascaded *RC* equaliser (refer to chapter 3 for theory of operation) which peaked at 15 MHz but had an overall bandwidth of approximately 100 MHz. The output of the equaliser fed the LED clusters via a current mirror and the entire transmitters frequency response was 10 MHz. This was selected purposely by the authors since 10BASE-T Ethernet requires a 10 Mb/s data rate and Manchester[1] line coding [5]. The authors wanted to maintain a realistic link, which meant also adhering to the minimum International Organisation for Standardisation (ISO) standards for office illumination, which stand at 350 lx [20], while operating over a

[1] Incidentally, the Manchester line code is named after the Manchester Mark I computer, where it was first used to store data at the university [19].

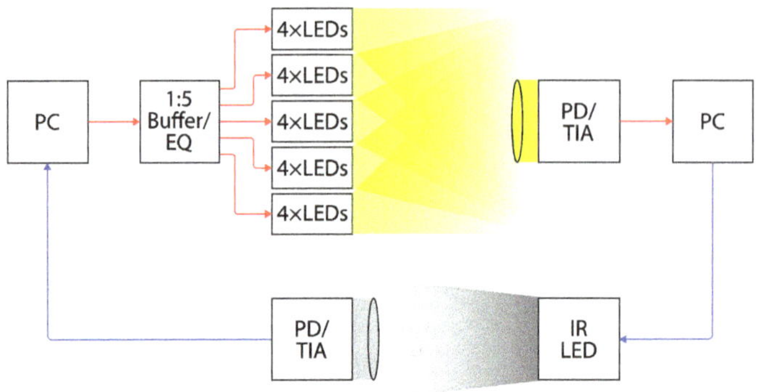

Figure 6.8. The system used to demonstrate an IR uplink for the first time in [16]; 20 LEDs were used in the downlink to ensure sufficient optical power reached the receiver, while a single IR-LED was used in the uplink.

2 m distance, which is typical of ceiling-to-table distances in an office environment. The test was performed in a small room simulator, that was ~1.4 × 1.7 × 2 m (length × width × height). Thus, the resistive equaliser could not extend the bandwidth further than 10 MHz, taking into account all of these considerations. The transmitter circuit that was used is illustrated in figure 6.9, along with the components and values used in the report.

The uplink was a more trivial exercise. One of the huge advantages that IR-based communication has over VLC is the inherently wider bandwidths due to lack of restriction from the slow phosphor-based colour-conversion used to obtain white light. The bandwidth of the single IR-LED used was 25 MHz and was collimated by a biconvex lens to focus the light on the receiver. A single device was used to maintain small form factor as the envisaged application is for mobile devices that cannot utilise large clusters of transmitters.

Among the results reported, it was shown that the minimum illuminance measured was 276 lx in the corners of the room, while the vast majority of the space was well illuminated. Regardless, even in the spaces where illumination failed the ISO standard, a bi-directional system supporting 10BASE-T Ethernet was demonstrated with an IR uplink and a maximum BER of 4.4×10^{-10}, indicating that there was excess SNR and additional resources were available to use. The packet transmission was performed in UDP and constant bit rate mode, which means that packets were continually sent in broadcasting mode with no link handshakes or error control. No detrimental performance was measured regardless of the packet size, and error free transmission rates were recorded up to 800 packets/second.

In the subsequent years, a number of developments occurred in IR uplinks and an interesting one was the addition of a TDMA-like interface for multiple users. Presuming the concept access network illustrated in figure 6.10, all users would require medium access whilst using the same wavelength, and all have visibility of every access point, the authors of [21] have proposed three separate TDMA configurations to enable multi-user connectivity, which was the first report of a smart multi-access method in any VLC network.

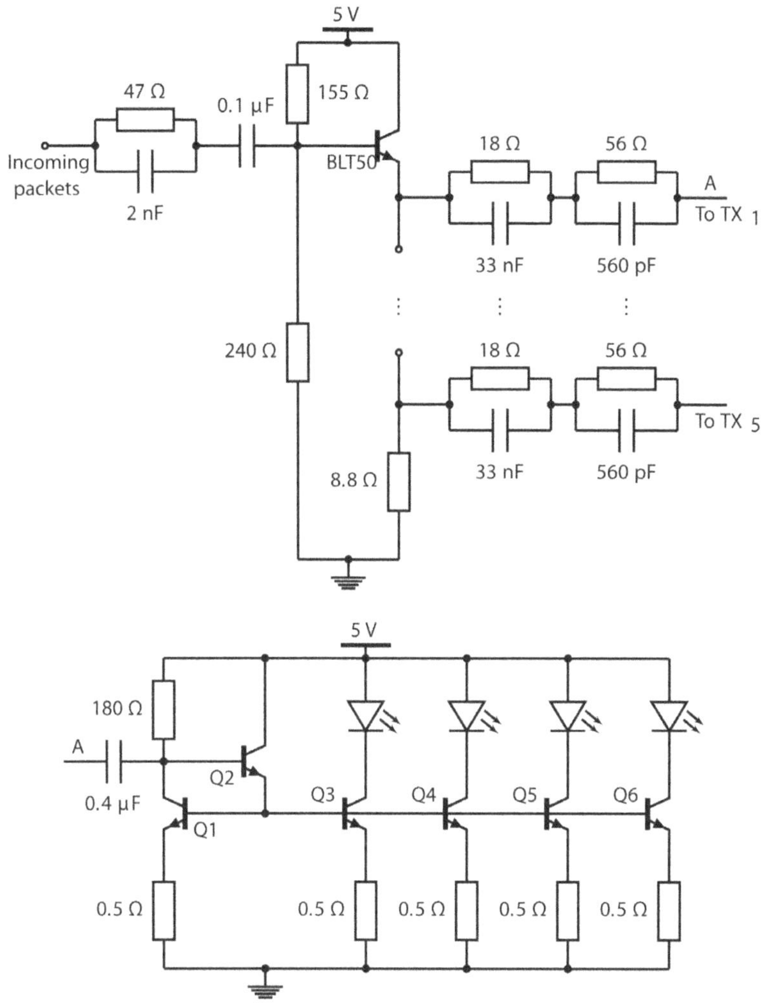

Figure 6.9. A circuit diagram of the transmitters used in [16].

In the report, the authors utilise the multi-face PD first proposed by Burton *et al* in [22–24] and illustrated in figure 6.11, also shown in chapter 3. This receiver enables full mobility since a signal can be received from any possible direction and variations of this receiver have been re-iterated numerous times for different applications in the literature [25–29].

In [22], several geometries for the multi-element receiver are presented, finally settling on a hexagonal structure as the best trade-off between coverage and design complexity. A proposal for the receiver schematic can be seen in figure 6.12 (software flowchart in figure 6.13), which consists of seven PDs due to the hexagonal structure, a TIA and limiting amplifier for each receiver. There is a microcontroller that reads the received signal strength indicator (RSSI) and makes a decision

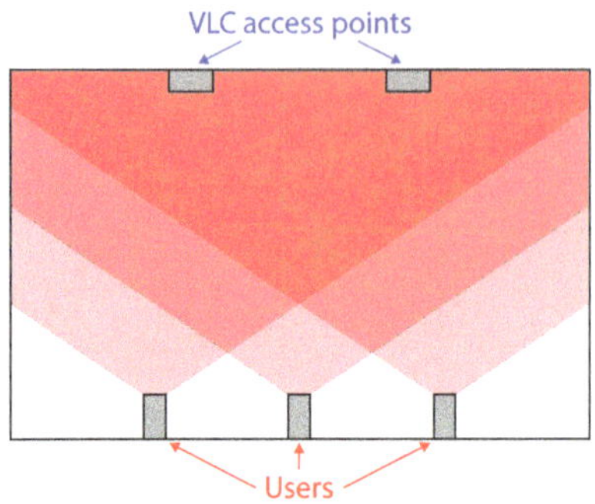

Figure 6.10. The generic overall architecture of a multi-user scenario, where multiple users are all transmitting simultaneously to interact with the VLC access points. The red colour is used illustratively to highlight the overlap between users and how they may have to compete to gain medium access, or be assigned specific resources that may puncture throughputs.

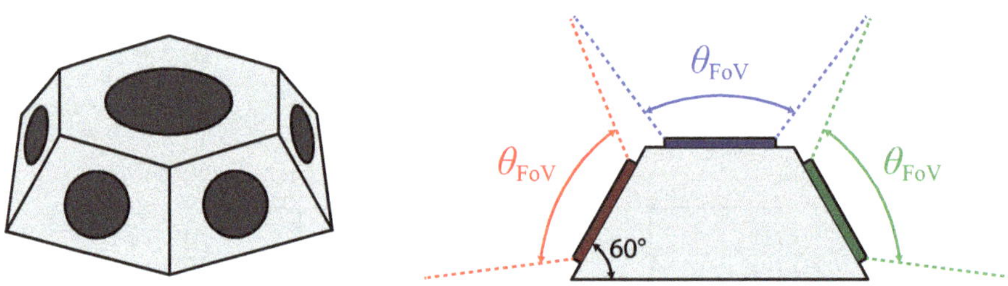

Figure 6.11. The multi-element receiver originally reported in [22–24] and later used in numerous works [25–29].

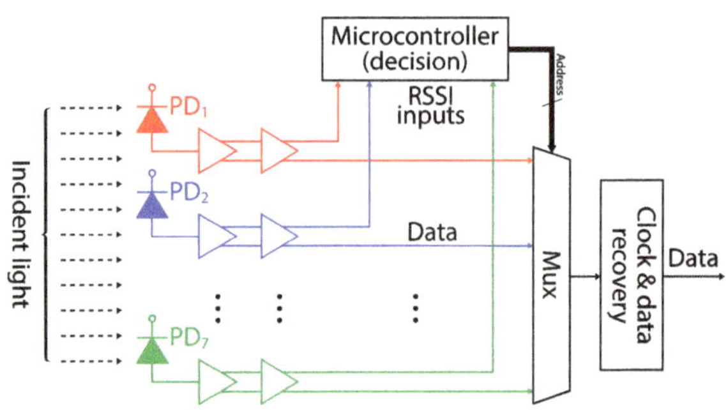

Figure 6.12. The schematic that defines the multi-element receiver operation. Each PD-TIA element pair receives a portion of the incident light and reports it to the microcontroller that makes a decision. The microcontroller then selects the signal path with the highest RSSI before onward detection and demodulation of the signal is performed.

regarding from which PD to receive the signal from. A multiplexer then selects that channel and demodulation is performed.

A summary of geometries and their associated FoV and the optimal structural angle can be found in table 6.1, where the number of sides listed does not include the top-facing side for the benefit of the tile angle, and the top-facing FoV is given by [22]:

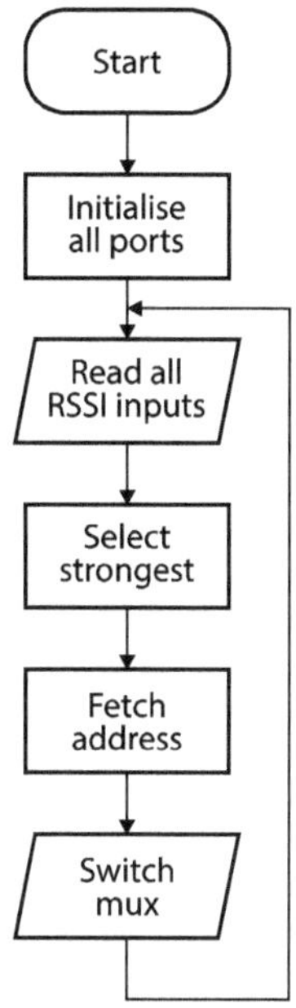

Figure 6.13. A flowchart that defines the operation of the microcontroller setup shown in the schematic in figure 6.12.

Table 6.1. The angular properties of the multi-element receiver.

No. sides	FoV_{sides} (°)	Tilt angle (°)	FoV_{top} (°)
3	120	60	N/A
4	90	45	N/A
5	72	36	36
6	60	30	60
7	51.43	25.715	77.14
8	45	22.5	90

$$FoV_{top} = 180° - 2FoV_{sides} \tag{6.1}$$

where FoV_{sides} is given in turn by [22]:

$$FoV_{top} = \frac{360°}{N_{sides}} \tag{6.2}$$

where N_{sides} is the number of side-facing sides and therefore the tilt angle α is given by [22]:

$$\alpha = \frac{FoV_{sides}}{2} \tag{6.3}$$

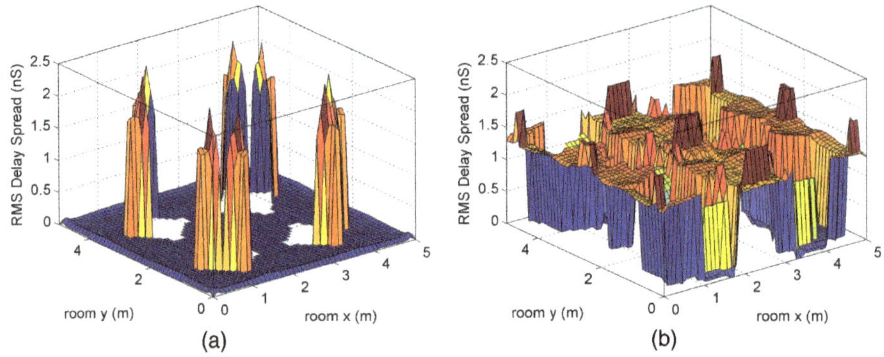

Figure 6.14. The RMS delay spread for (a) the system without the multi-element receiver and (b) with it. Although the multi-element receiver increases the overall delay spread on average, it does substantially enhance mobility. [22] John Wiley & Sons. © 2014 WILEY-VCH Verlag GmbH & Co., KGaA, Weinheim.

The receiver ultimately trades mobility, where a signal can be received from every position in the room against a slightly higher RMS delay spread, see figure 6.14 [22].

In [21], three TDMA-like slot structures were proposed to achieve multiple access, namely (i) conservative- (ii) pure- and (iii) face selective-TDMA. Illustrated examples of each scheme are respectively shown in figures 6.15(a)–(c), where each of the schemes are illustrated in conceptual form. For figure 6.15(a), the slots are pre-defined for all access points and users, while for (b), the strongest signal component is the one considered, and finally for (c), there is a third dimension, which is the specific detector that is used to recover the signal, and these can be independent for different links.

For (i) the idea is to minimise the potential overlaps between users, isolating them as much as possible. This will result in a lack of interference and is achieved by guaranteeing the user a regular time-slot for each access point on the network, regardless of the resource request. This has the clear advantage that each user has an equal opportunity to transmit, however, the power received by each access point is not equal due to the distance, and hence, some resources are wasted. The maximum number of users N_U that can be supported in this scheme is given by [21]:

$$N_U = \left\lfloor \frac{\mathcal{R}}{\mathcal{R}_u} \right\rfloor \tag{6.4}$$

where $\mathcal{R}$ and $\mathcal{R}_u$ are the aggregated rate of the link and the minimum rate provided to each user, respectively. When using the multi-element receiver outlined above, maximal-ratio combining (MRC) can be used to select the strongest signal component from each of the PDs.

The signal-to-interference-plus-noise ratio (SINR) can therefore be calculated according to the number of access points N_A in the network and the number of faces N_f present in the multi-element receiver. It is defined as [21]:

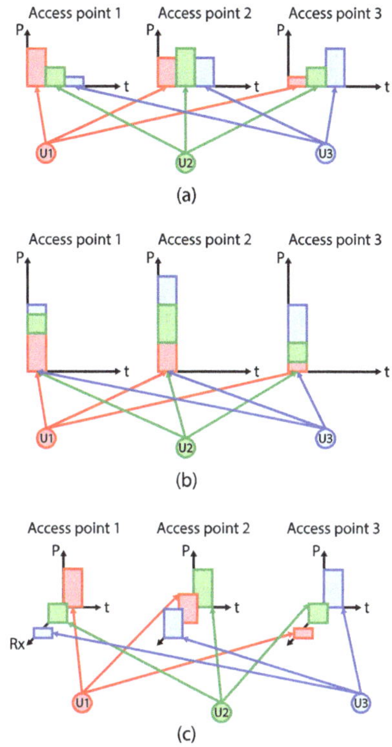

Figure 6.15. The three TDMA-like slot structures presented in [21], including (a) conservative-TDMA, pure-TDMA and (c) face selective-TDMA.

$$SINR = \sum_{j=1}^{N_A} \sum_{k_j=1}^{N_F^{(j)}} \beta_{k_j,j} \tag{6.5}$$

where the term j defines the current user, so $N_f^{(j)}$ is the number of faces that the jth user can see. The term that is summed is the reference SNR given in turn by [21]:

$$\beta_{k_j,j} = \frac{P_t \left| h_{ij}^{(k_j)} \right|^2}{N_0 B} \tag{6.6}$$

where B is the signal bandwidth and N_0 is the noise spectral density. Equation (6.6) can clearly be linked to (3.17) in chapter 3. It becomes relatively straightforward to see the disadvantages of this scheme, since different access points will receive different power levels from each of the users and hence, decoding the information becomes heavily dependent on the weakest SINR received, which is a problem. Furthermore, according to (6.4), the number of users limits the throughput that can be supported.

As a result, scheme (ii) was introduced, the so-called pure-TDMA, which accepts interference from different sources, removing the fixed slot allocation-per-user, improving throughput. Instead, all users transmit simultaneously and it is always presumed that one access point has a link with a single dominant user and the remaining users are simply low-grade interferers. Therefore, the maximum number of users able to connect is updated as follows [21]:

$$\bar{N}_U = N_A \left\lfloor \frac{\mathcal{R}}{\mathcal{R}_u} \right\rfloor \tag{6.7}$$

recalling that N_A is the number of access points in the network. When considering MRC, the SINR is given by [21]:

$$\beta_{k_j,j} = \frac{P_t \left| h_{ij}^{(k_j)} \right|^2}{N_0 B + \sum_{\ell=1}^{N_A - 1} \left| h_{\ell j}^{(k_j)} \right|^2} \tag{6.8}$$

where the term $\sum_{\ell=1}^{N_A-1} \left| h_{\ell j}^{(k_j)} \right|^2$ is introduced, which describes the contributions of the $N_A - 1$ interferers that are communicating when the signal of interest is received and $h_{lj}^{k_j}$ describes the channel gain between the lth transmitter and the k_jth face of the jth access point.

Obviously, having interferers is not ideal, since if an interfering transmitter is located an equal distance away from the access point as the signal of interest, the net SINR is reduced substantially, making the signal unrecoverable. Therefore, face selective-TDMA was proposed in [21] that adds an additional dimension to the scheme, which is spatial diversity.

This scheme effectively uses the multiple faces of the receiver to diversify the LOS paths between transmitters and access points. If there are two signals incident on different faces of the receiver with two different LOS paths, then it is possible to add DSP complexity to separate the signals and demodulate them separately.

Therefore, the maximum number of users can once again be updated to the following [21]:

$$\bar{\bar{N}}_U = \alpha N_A \left\lfloor \frac{\mathcal{R}}{\mathcal{R}_u} \right\rfloor \tag{6.9}$$

and the term α is a number where $\alpha = [1/N_A, N_F]$ where N_F is the number of faces in total on the receiver, while the SINR is now given by:

$$\beta_{k_j,j} = \frac{P_t \gamma_{k_j} \left| h_{ij}^{(k_j)} \right|^2}{N_0 B + \sum_{\ell=1}^{N_A - 1} \gamma_{k_j} \left| h_{\ell j}^{(k_j)} \right|^2} \tag{6.10}$$

where the γ_{k_j} term is added to represent interference on the k_jth face.

The authors of [21] numerically simulated a system with a receiver that had $N_F = 5$ faces. A number of performance indicators were measured for each TDMA scheme, most importantly including the average SINR and the maximum number of users as a function of the number of access points present in a given room. The room dimensions tested by the authors were 9×9 (length $\times$ width), however, the height of the room is not listed in the work. The total transmitted power of the LEDs is 1 W and the data rate is set to 10 Mb/s.

The maximum number of users that can be supported by each scheme is illustrated in figure 6.16. Clearly, for conservative-TDMA, due to the fact that fixed time-slots are allocated, the number of users is hard limited by (6.4) and not variable. The pure-TDMA scheme offers improved performance, since it relies on a single, high power LOS link between a specific user and the face of an access point, and other users are treated as low-grade interference noise. Therefore, the number of users can improve with the number of access points, since the number of available faces also increases linearly with the number of access points. As expected, however, face selective-TDMA clearly offers the best performance since it takes advantage of the diversity of both time and space. The improvement offered by face selective-TDMA improves with the number of access points in a similar manner to pure-TDMA, but at a higher rate. When $N_A = 5$, 35 users can be supported for face selective-TDMA in comparison to 20 and 4 for the pure- and conservative-TDMA schemes, illustrating the vast improvement offered by the face selective approach in terms of the number of users connected.

On the other hand, the number of users able to connect does not necessarily imply improved performance, as illustrated in figure 6.17. The best performance is constantly offered by conservative-TDMA, which offers the best SINR in every case and it increases with N_A since more access points can recover the signal and combine them together. Clearly, since pure-TDMA assumes that the signal with the

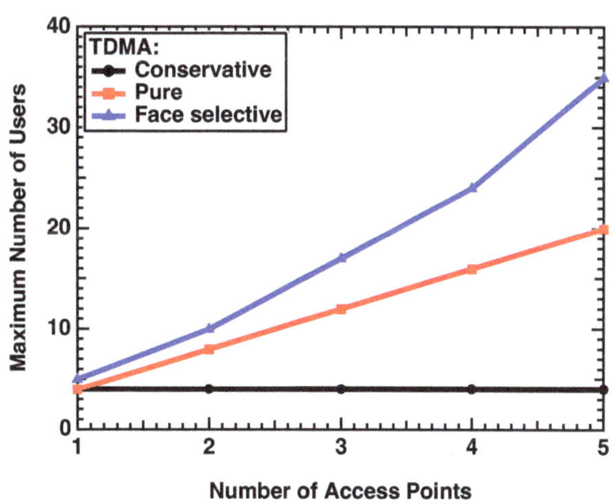

Figure 6.16. The number of users that can be supported with each TDMA-like slot structure proposed in [21]. Due to the additional degree of freedom introduced by face selective-TDMA, a higher number of users can be supported, which increases with the number of access points available.

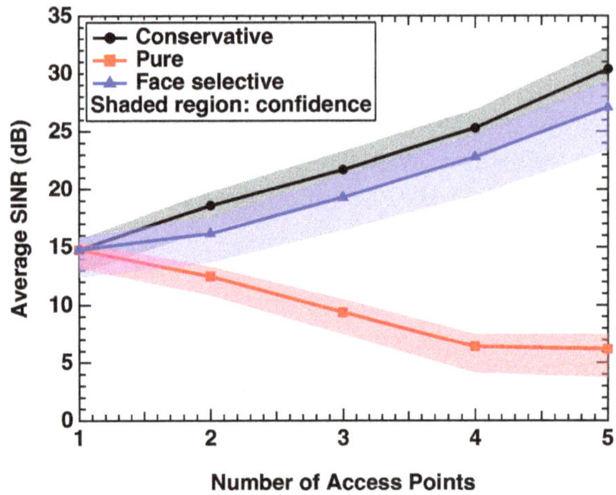

Figure 6.17. The average SINR measured for each TDMA-like system with respective confidence intervals. The only system that shows a performance drop with the number of access points is pure-TDMA, because there is increased interference between access points.

highest power is the one recovered, as the number of access points increases, there can be more interference and hence, the SINR is reduced. The face selective-TDMA scheme offers a slightly reduced SINR in comparison to the conservative scheme since it allows overlap of the transmissions; however, allowing the different faces of the receivers to work independently compensates for the reduction of SINR exhibited in the pure-TDMA case. Therefore, overall, the face selective scheme allows optimal use of resources in a TDMA setting, and shows efficient use of a IR uplink.

6.1.3 Hybrid radio frequency systems

Possibly the most obvious of all the possible uplink candidates is RF for compatibility with current technologies and mobile devices. The report in [30] argues that wide-angle aperture-based optics are not well suited to uplinks due to their form factor and energy constraints, which is a compelling argument in the opinion of the author.

They therefore realise a light-radio (LiRa) wireless network that operates at the medium access protocol (MAC) layer. To implement this, a MAC system that controls the access for both the Wi-Fi and VLC systems simultaneously was developed. The operating principles of this MAC are as follows; first the access point performs a VLC automatic repeat request. To avoid blocking the generic Wi-Fi traffic, the IEEE 802.11 data-acknowledge patterns are blocked and instead a controlled trigger message is spoofed with an appropriate network allocation vector that is selected purposely in excess to allow multiple clients to transmit, and the appropriate order is also included in the message to avoid contention. Finally, the access point sends the trigger that is used to balance airtime.

The LiRa link was developed for the Mango Communications WARP v3 software-defined radio (SDR) [30], which is a real time development board based on a Virtex 6 field programmable gate array (FPGA). It has 40 MHz bandwidth and is also capable of multiple-input multiple-output (MIMO) clocking. The test scenario built by the authors of [30] is illustrated in figure 6.18. The LiRa architecture is connected to plural VLC transmitters and a Wi-Fi access point. The LEDs can be either used in combination to support mobility, or they can be used individually to support individual users and avoid interference. Each of the client devices are equipped with at least one PD but the LiRa router is not, since there is no optical uplink. Where possible, plural PDs are used on multiple faces of the receiver surface to introduce angle diversity as described in the previous section. All downlink traffic is routed through the VLC link while the RF handles the uplink. One advantage of the LiRa router is that it maintains compatibility for non-VLC enabled devices, and can support bi-directional traffic if such a device connects to the network.

The reason an optical uplink is not included in [30] was because including a transmit array of LEDs is undesirable due to the reduced transmit power requirements and illumination intensity, meaning that sufficient rates could not be achieved [31, 32]. Furthermore, infrared uplinks are subject to rotational fades that result in outages, if the angle of rotation is beyond 15° [30].

The software stack that LiRa uses is a custom developed modification of the IEEE 802.2 interface [33]. The conventional IEEE 802.2 logical-link control layer is used with a separate MAC for the LiRa interface and legacy IEEE 802.11 [34] technologies. After aggregation, at the LiRa access point there are separate physical layers for VLC and the legacy connections. Transmission occurs over the technology-specific channel before reception by a PD (LiRa) or antenna (legacy). The reverse of the above makes up the receiver. This entire stack is shown in figure 6.19. The physical layer is not fixed in the proposed LiRa and can consist of any

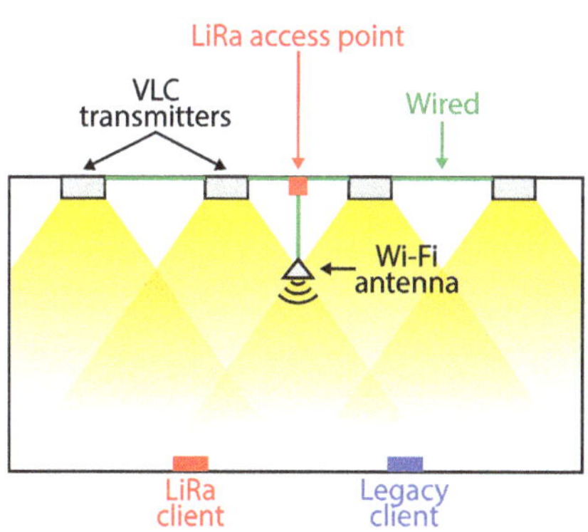

Figure 6.18. The scenario presented in the report [30] is illustrated. The LiRa system can tolerate transmission with both LiRa clients that have both VLC and Wi-Fi capabilities and legacy clients which only have Wi-Fi antenna. The VLC transmitters are spread out over the ceiling to ensure mobility and hand-off between access points can continually provide both connection technologies to the user.

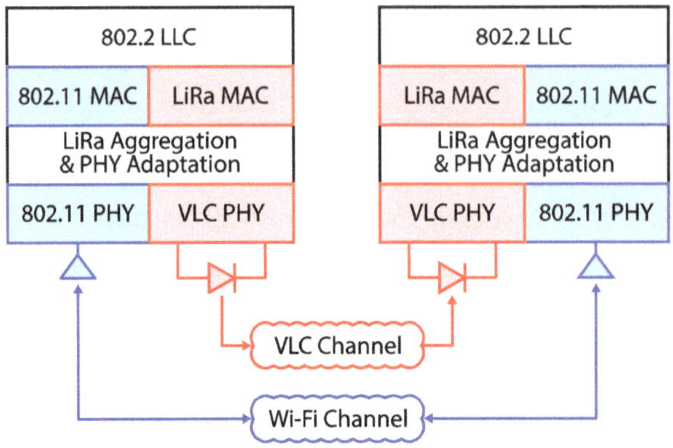

Figure 6.19. The hybrid stack proposed by the LiRa approach in [30]. The two MACs operate independently and feed the aggregation layers (or vice-versa) that allocate the transmission flows to each candidate technology according to demand and availability.

technology reported in the literature, including (but not limited to) the highest speed links available at the time of writing [35–37].

Around the same time, the authors of [38–41] illustrated a similar concept and one of the most highly cited articles that investigates a fully bi-directional system was reported in [38]. It is argued that hybrid networks utilising heterogeneous networks will improve QoS in the indoor environment where the transmission distance is short range. Advantages of embedding multiple technologies mainly focus on the ability to combine small-cells with atto-cells provided by different technologies. For example, indoor micro-cells could be provided by Wi-Fi to ensure mobility and continuous connection while a user moves either between VLC atto-cells or between rooms. Furthermore, loading of traffic between technologies will be enabled, thus reducing the stress on the available resources. It is also argued that security is improved in heterogenous networks in [38], since optical signals cannot penetrate walls[2].

Since the vast majority of internet traffic is generated and received indoors (up to 80% [42]), additional throughput will certainly be required as time moves forward, there is a palpable space for heterogeneous networks that can provide extra resources, which is in line with existing technologies as modern mobile devices are generally tri-/quad-band out-of-the-box.

A demonstrator is presented across the work reported in [38–41] where a co-operative Wi-Fi-VLC network that uses the two topologies is illustrated in figures 6.20(a) and (b), which show hybrid and co-operative versions of a link. The key difference between the two is that VLC and Wi-Fi are used exclusively for the downlink and uplink, respectively, in figure 6.20(a), while a bi-directional VLC and Wi-Fi links are considered in figure 6.20(b).

[2] This is a subjective statement in the opinion of the author, since simultaneous connections to multiple access points introduces additional vulnerabilities not present in homogeneous networks.

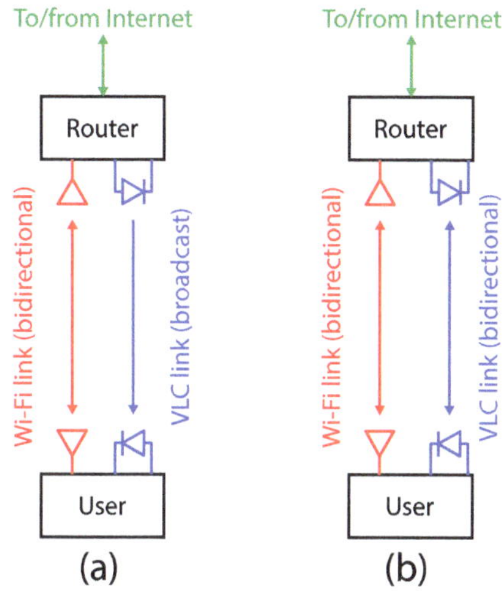

Figure 6.20. The two different hybrid architectures presented in [38–41]. Note that the Wi-Fi link is always bi-directional, however, in (a) the VLC link is broadcast-only, while in (b) it is bi-directional.

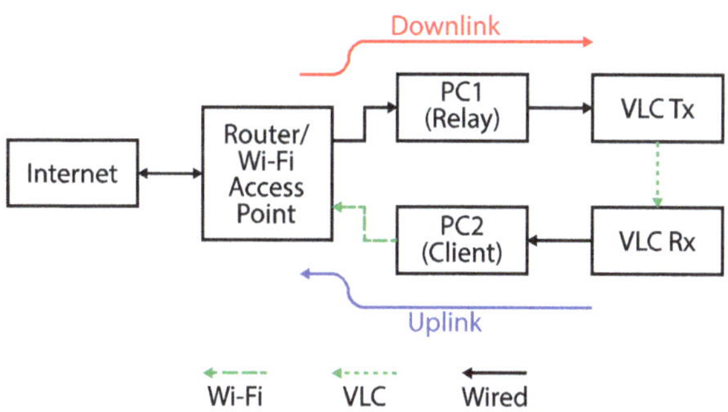

Figure 6.21. The architecture of the hybrid link.

The architecture of the hybrid system presented in [38] is illustrated in figure 6.21. The internet backbone is connected into a standard Wi-Fi router that connects to a PC via a standard wired Ethernet connection and another PC via a Wi-Fi link as the uplink [41]. On the other hand, the co-operative architecture is illustrated in figure 6.22 and connects the router with single a PC via standard Wi-Fi but also through a series of VLC transceivers.

The system presented is a realistic implementation of such a link, with realistic data rates and implementation of the open systems interconnection (OSI) stack that includes layers 1 and 2. The LED used was driven by an analogue pre-equaliser to improve the bandwidth of the link to 180 MHz. Focusing optics were used to narrow the beam and improve the power impinging on the detector at the receiver.

The VLC signals were designed to convert ethernet packets via a media converter to 70 MHz OFDM signals complete with pilot tones for frequency-domain equalisation at the receiver. After such equalisation, channel state information is provided to the transmitter to bit-load the individual symbols and provide a link that is

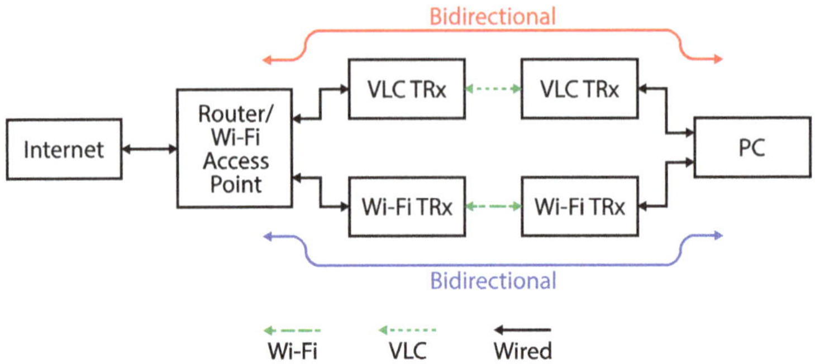

Figure 6.22. The architecture of the co-operative link.

optimised as far as possible. This leads to a maximum data rate of 500 Mb/s, depending on the quality of transmission with a latency of around 10 ms. The Wi-Fi standard used was IEEE 802.11n set to 54 Mb/s mode.

The second OSI layer contains a MAC that controls when a user can access the resources of the technology used to transmit information. These are usually defined by standards focused on utilising a single technology. In [30, 39], new MACs were developed to deal with multiple technologies and this is something that should be adopted in beyond 5G technologies using the general approach illustrated in figure 6.19. The MAC should optimise the traffic flows through the different technologies based on resource availability and user requirements.

Each system was tested as a function of distance and blocking time-per-minute. Clearly, the Wi-Fi throughput is constant at ~30 Mb/s regardless of distance and this would be expected due to the mobility enabled through Wi-Fi. Interestingly, the hybrid system that utilises VLC exclusively as a downlink offers an improved performance for a distance ⩽4 m and actually degrades performance for higher distances. The reason for this is attributed to the fact that the VLC link has an inverse square law with distance. The co-operative solution offers the best performance, as expected, since as VLC speeds diminish due to distance, Wi-Fi can take over as the dominant link. The measured performance of the throughput as a function of blocking is shown in figure 6.23.

In terms of throughput when blocking is considered, both systems outperform the Wi-Fi-only link. Blocking is an important parameter to consider because due to the movement of people, the LOS between transmitter and receiver may occasionally be blocked. Even when the LOS link is blocked up to half the time (30 s/minute), the performance of the hybrid and co-operative links are superior to Wi-Fi by itself.

Tests were performed that investigated the average load pages of a series of popular websites, and it was ubiquitously shown that the hybrid system shows the best performance, loading the popular sites the fastest, followed by the co-operative system and then the Wi-Fi only system, with the results illustrated in figure 6.24.

This may seem counter-intuitive because the co-operative system has more resources, however, making a decision on which link to use and then routing the signals appropriately adds additional latency to the link. The authors of [39]

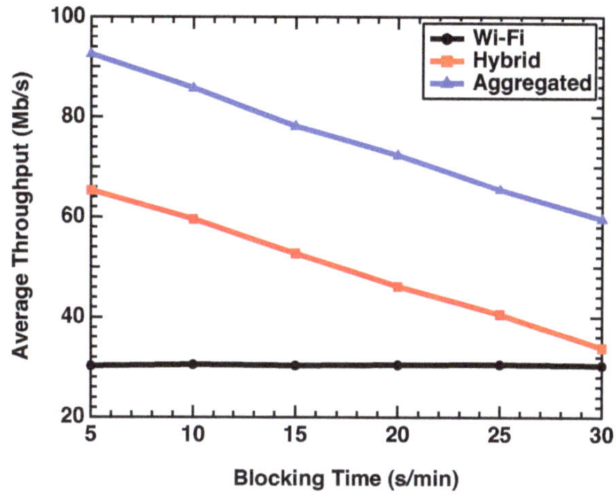

Figure 6.23. The measured average throughput of the hybrid and co-operative system in comparison to Wi-Fi as a function of blocking time. Both systems comfortably outperform Wi-Fi due to being able to offer an increased number of traffic flows at higher rates due to additional technological resources [38–41].

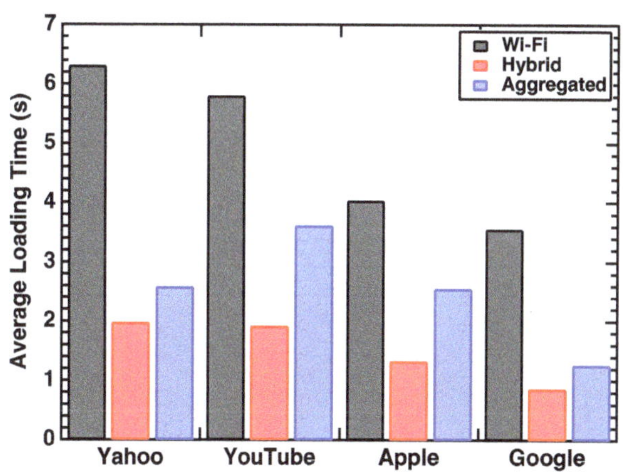

Figure 6.24. The average loading times of each system for a number of popular websites and companies [38–41].

illustrated this succinctly via a histogram of the load-times for one of the web pages over thirty separate site requests, see figure 6.25. The co-operative and hybrid system average at 3.56 and 1.97 s, respectively, while the Wi-Fi system averages at over 6 s.

6.2 Summary

In this chapter, uplink technologies were discussed. The question of uplink technologies is an interesting one in the opinion of the author. It is a topic that has rarely been discussed with any significant depth but remains possibly the single most important question remaining for VLC as an access network technology for home and office use. If a common uplink cannot be found then proliferation of VLC

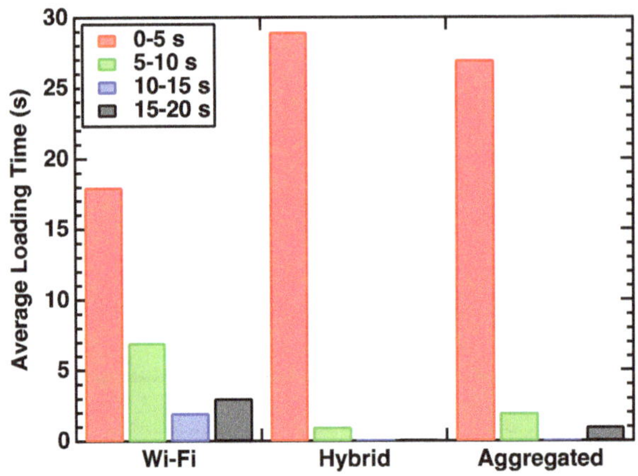

Figure 6.25. The average loading time histogram for each system, which clearly shows that the hybrid approach offers superior performance and the lowest loading times, while the aggregated system also outperforms the Wi-Fi-only system [38–41].

systems will be troublesome. For example, commercial vendors have proposed the use of IR as the uplink, which is problematic because it has advantages over visible technologies that would work reciprocally in the downlink.

Therefore, this chapter has discussed the possible solutions for the uplink in a VLC system, with particular focus on the hybrid approach, which in the opinion of the author is the most promising. The author strongly believes that hybrid and heterogeneous technological approaches to next generation networks are the most favourable option and there is a large gap in the market, both academically and industrially, for implementations of hybrid MACs that can make the best use of each technology, where the demand is required.

References

[1] Holma H, Toskala A and Nakamura T 2020 *5G Technology: 3GPP New Radio* (New York: Wiley)

[2] Haigh P A, Son T T, Bentley E, Ghassemlooy Z, Le Minh H and Chao L 2012 Development of a visible light communications system for optical wireless local area networks *2012 Computing, Communications and Applications Conf.* (IEEE) pp 351–5

[3] Postel J 1981 User datagram protocol. RFC768 *Internet Control Message Protocol-DARPA Internet Program Protocol Specification-RFC* **791** 109–13

[4] Postel J 1981 Transmission control protocol. RFC793 *Internet Control Message Protocol-DARPA Internet Program Protocol Specification-RFC*

[5] Spurgeon C E 2000 *Ethernet: The Definitive Guide* (Senastopol, CA: O'Reilly Media, Inc.)

[6] Gomez A, Shi K, Quintana C, Faulkner G, Thomsen B C and O'Brien D 2016 A 50 Gb/s transparent indoor optical wireless communications link with an integrated localization and tracking system *J. Lightwave Technol.* **34** 2510–7

[7] Gomez A, Shi K, Quintana C, Maher R, Faulkner G, Bayvel P, Thomsen B C and O'Brien D 2016 Design and demonstration of a 400 Gb/s indoor optical wireless communications link *J. Lightwave Technol.* **34** 5332–9

[8] Beyranvand H, Lévesque M, Maier M, Salehi J A, Verikoukis C and Tipper D 2016 Toward 5G: FiWi enhanced LTE-A HetNets with reliable low-latency fiber backhaul sharing and WiFi offloading *IEEE/ACM Trans. Netw.* **25** 690–707

[9] Zhang H, Wang R, Wang H and Wang Q 2019 Backup fibre deployment algorithm based on daily traffic demand in fibre-wireless (FiWi) access networks *IET Commun.* **13** 2956–65

[10] Wang Y, Chi N, Wang Y, Tao L and Shi J 2015 Network architecture of a high-speed visible light communication local area network *IEEE Photonics Technol. Lett.* **27** 197–200

[11] Lu F, Cheng L, Shi J, Xu M, Wang J, Shen S and Chang G-K 2017 Efficient mobile fronthaul incorporating VLC links for coordinated densified cells *IEEE Photonics Technol. Lett.* **29** 1059–62

[12] Shi J, Hong Y, Deng R, He J and Chen L-K 2020 Real-time software-reconfigurable hybrid in-house access with OFDM-NOMA *IEEE Photonics Technol. Lett.* **32** 379–82

[13] Gupta A, Sharma N, Garg P and Alouini M-S 2017 Cascaded FSO-VLC communication system *IEEE Wireless Commun. Lett.* **6** 810–3

[14] Huang Z, Wang Z, Huang M, Li W, Lin T, He P and Ji Y 2017 Hybrid optical wireless network for future sago-integrated communication based on FSO/VLC heterogeneous interconnection *IEEE Photonics J.* **9** 1–10

[15] Pesek P, Zvánovec S, Chvojka P, Ghassemlooy Z and Haigh P A 2018 Demonstration of a hybrid FSO/VLC link for the last mile and last meter networks *IEEE Photonics J.* **11** 1–7

[16] Burton A, Bentley E, Le Minh H, Ghassemlooy Z, Aslam N and Liaw S-K 2014 Experimental demonstration of a 10BASE-T Ethernet visible light communications system using white phosphor light-emitting diodes *IET Circ. Devices Syst.* **8** 322–30

[17] Khalid A M, Cossu G, Corsini R, Choudhury P and Ciaramella E 2012 1-Gb/s transmission over a phosphorescent white LED by using rate-adaptive discrete multitone modulation *IEEE Photonics J.* **4** 1465–73

[18] Kassem A and Darwazeh I 2019 Exploiting negative impedance converters to extend the bandwidth of LEDs for visible light communication *2019 26th IEEE Int. Conf. on Electronics, Circuits and Systems (ICECS)* pp 302–5

[19] Lavington S H 1978 The Manchester Mark I and atlas: a historical perspective *Commun. ACM* **21** 4–12

[20] Burton A, Le Minh H, Ghassemlooy Z, Bentley E and Botella C 2014 Experimental demonstration of 50-Mb/s visible light communications using 4 × 4 MIMO *IEEE Photonics Technol. Lett.* **26** 945–8

[21] Bui T-C and Biagi M 2019 TDMA-like infrared uplink with multi-faces photodiode access points *2019 IEEE Int. Conf. on Communications Workshops (ICC Workshops)* (IEEE) pp 1–5

[22] Burton A, Ghassemlooy Z, Rajbhandari S and Liaw S-K 2014 Design and analysis of an angular-segmented full-mobility visible light communications receiver *Trans. Emerg. Telecommun. Technol.* **25** 591–9

[23] Burton A, Le Minh H, Ghassemlooy Z, Rajbhandari S and Haigh P A 2012 Performance analysis for 180° receiver in visible light communications *2012 4th Int. Conf. on Communications and Electronics (ICCE)* (IEEE) pp 48–53

[24] Burton A, Le Minh H, Ghassemlooy Z, Rajbhandari S and Haigh P A 2012 Smart receiver for visible light communications: design and analysis *2012 8th Int. Symp. on Communication Systems, Networks & Digital Signal Processing (CSNDSP)* (IEEE) pp 1–5

[25] Chen C, Bian R and Haas H 2018 Omnidirectional transmitter and receiver design for wireless infrared uplink transmission in LiFi *2018 IEEE Int. Conf. on Communications Workshops (ICC Workshops)* (IEEE) pp 1–6

[26] Chen C, Zhong W-D, Yang H, Zhang S and Du P 2018 Reduction of SINR fluctuation in indoor multi-cell VLC systems using optimized angle diversity receiver *J. Lightwave Technol.* **36** 3603–10

[27] Chen C, Zhong W-D, Wu D and Ghassemlooy Z 2016 Wide-FOV and high-gain imaging angle diversity receiver for indoor SDM-VLC systems *IEEE Photonics Technol. Lett.* **28** 2078–81

[28] Zhong W-D, Chen C, Yang H and Du P 2017 Performance analysis of angle diversity multi-element receiver in indoor multi-cell visible light communication systems *2017 19th Int. Conf. on Transparent Optical Networks (ICTON)* (IEEE) pp 1–4

[29] Fahamuel P, Thompson J and Haas H 2014 Improved indoor VLC MIMO channel capacity using mobile receiver with angular diversity detectors *2014 IEEE Global Communications Conf.* (IEEE) pp 2060–5

[30] Naribole S, Chen S, Heng E and Knightly E 2017 LiRa: A WLAN architecture for visible light communication with a Wi-Fi uplink *2017 14th Annual IEEE Int. Conf. on Sensing, Communication, and Networking (SECON)* (IEEE) pp 1–9

[31] Pathak P H, Feng X, Hu P and Mohapatra P 2015 Visible light communication, networking, and sensing: a survey, potential and challenges *IEEE Commun. Surv. Tutor.* **17** 2047–77

[32] Noshad M and Brandt-Pearce M 2013 Can visible light communications provide Gb/s service? arXiv:1308.3217

[33] IEEE 1989 IEEE standard for information technology—telecommunications and information exchange between systems—local and metropolitan area networks—specific requirements—part 2: Logical link control *ISO 8802-2 IEEE 802.2, First Edition 1989-12-31 (Revision of IEEE Std 802.2-1985)* pp 1–114

[34] IEEE 2016 IEEE standard for information technology-telecommunications and information exchange between systems local and metropolitan area networks-specific requirements—part 11: wireless LAN medium access control (MAC) and physical layer (PHY) specifications *IEEE Std 802.11-2016 (Revision of IEEE Std 802.11-2012)* pp 1–3534

[35] Bian R, Tavakkolnia I and Haas H 2019 15.73 Gb/s visible light communication with off-the-shelf LEDs *J. Lightwave Technol.* **37** 2418–24

[36] Fahs B, Chowdhury A J and Hella M M 2016 A 12-m 2.5-Gb/s lighting compatible integrated receiver for OOK visible light communication links *J. Lightwave Technol.* **34** 3768–75

[37] Chun H, Gomez A, Quintana C, Zhang W, Faulkner G and O'Brien D 2019 A wide-area coverage 35Gb/s visible light communications link for indoor wireless applications *Sci. Rep.* **9** 1–8

[38] Ayyash M, Elgala H, Khreishah A, Jungnickel V, Little T, Shao S, Rahaim M, Schulz D, Hilt J and Freund R 2016 Coexistence of WiFi and LiFi toward 5G: concepts, opportunities, and challenges *IEEE Commun. Mag.* **54** 64–71

[39] Shao S, Khreishah A, Ayyash M, Rahaim M B, Elgala H, Jungnickel V, Schulz D, Little T D C, Hilt J and Freund R 2015 Design and analysis of a visible-light-communication enhanced WiFi system *J. Opt. Commun. Netw.* **7** 960–73

[40] Shao S, Khreishah A, Rahaim M B, Elgala H, Ayyash M, Little T D C and Wu J 2014 An indoor hybrid WiFi-VLC internet access system *2014 IEEE 11th Int. Conf. on Mobile Ad Hoc and Sensor Systems* (IEEE) pp 569–74

[41] Li Z, Shao S, Khreishah A, Ayyash M, Abdalla I, Elgala H, Rahaim M and Little T 2018 Design and implementation of a hybrid RF-VLC system with bandwidth aggregation *2018 14th Int. Wireless Communications & Mobile Computing Conf. (IWCMC)* (IEEE) pp 194–200

[42] Cisco 2019 Cisco visual networking index: global mobile data traffic forecast update, 2017–2022 (https://davidellis.ca/wp-content/uploads/2019/12/cisco-vni-mobile-data-traffic-feb-2019.pdf)

Chapter 7

Localisation and positioning

7.1 Introduction

One of the most important and useful developments of the 20th century, global navigation satellite systems (GNSSs) have revolutionised the way we live our lives. Traditionally, GNSSs are satellite-based navigation systems based on radio trilateration [1, 2] and the most commonly known is the global positioning system (GPS), which has penetrated our day-to-day activities, powering data-driven maps since its creation in 1973. Owned by the United States government and operated by the United States Space Force, GPS was originally developed as a military tool, until a civilian version was developed in the 1980s. Modernisation had continually been deployed in the 2000s, with an agreement signed between the United States government and the European Community to establish cooperation between the GPS network and Europe's similar Galileo project [3]. At the same time, Qualcomm had demonstrated successful tests of receiving GPS signals in mobile phones [4][1].

However, it is not without drawbacks and trade-offs and using radio waves leads to several significant challenges. One of the main challenges considers indoor applications; satellites can generally and accurately provide positioning within a high degree of accuracy outdoors [6, 7], however, satellites generally have poor material penetration and hence indoor coverage is highly limited [8]. This challenge is exacerbated when considering the impact of multi-path propagation in the indoor environment of highly attenuated signals [9–11]. Positioning using Wi-Fi has been attempted [12–14] but accuracy is generally lacking at an acceptable price point [15].

Recently, researchers have been utilising LEDs to develop indoor positioning capabilities that coexist with VLC systems [16–20]. The reason for this is because LEDs are generally placed in matrices in any given room, covering the ceiling with spotlights, and hence, trilateration can easily be achieved. This chapter discusses the

[1] It has become so fundamental to our daily lives that as recently as 2019, almost 50 years after its inception, it won the Queen Elizabeth Prize for Engineering worth £1M [5], indicating the lasting impact that GPS has had.

doi:10.1088/978-0-7503-1680-4ch7

state-of-the-art of such systems and their underlying mechanics that are capable of providing indoor positioning accuracy to sub-centimetre resolutions.

7.2 Positioning with visible light

Similar to data communication, the 'last-metre' problem is also a problem for positioning systems, mainly due to the multi-path impact on radio and attenuation of signals. Systems have been developed using Wi-Fi [21], bluetooth [22, 23] and several other short range personal area networks, with mixed degrees of success [24–26].

As VLC has emerged, the LEDs used have become attractive for accurate, low-cost positioning and localisation applications due to their advantages over RF antennas. The advantages are reported throughout the chapters of this book but include high bandwidths [27], low power consumption [28], long mechanical lifetimes [29] and cost efficiency [30].

Numerous schemes that utilise LEDs for positioning and localisation have subsequently been reported over the years, starting with, to the best of the author's knowledge, an innovation by the Japan Electronics and Information Technology Industries Association in 2007 [30] that proposed to standardise signals transmitted by LEDs that can lead to identification.

The general concept for a positioning system where the LEDs are in a configuration of known positions is illustrated in figure 7.1. Each LED takes turns to transmit a sinusoid or square wave with a given frequency in a predetermined time slot duration and frequency (see figures 7.2 and 7.3), employing both time and frequency multiplexing, ensuring that no signals ever overlap [31–34].

Over the years there have been numerous developments [35] using a suite of different technologies and several of them will be covered here. The first method to be covered is received signal strength (RSS), which includes the traditional trilateration approach.

7.2.1 Received signal strength

Trilateration

Trilateration is the most commonly found form of RSS-based localisation and this method has been demonstrated in numerous VLC systems [35–38] and it requires three sources placed at known positions. Considering only the LOS path, the

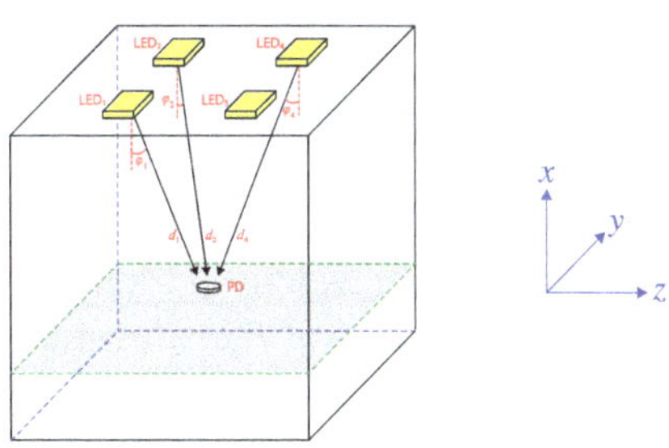

Figure 7.1. Positioning scenarios require at least three transmitters to be able to communicate with the receiver to enable trilateration.

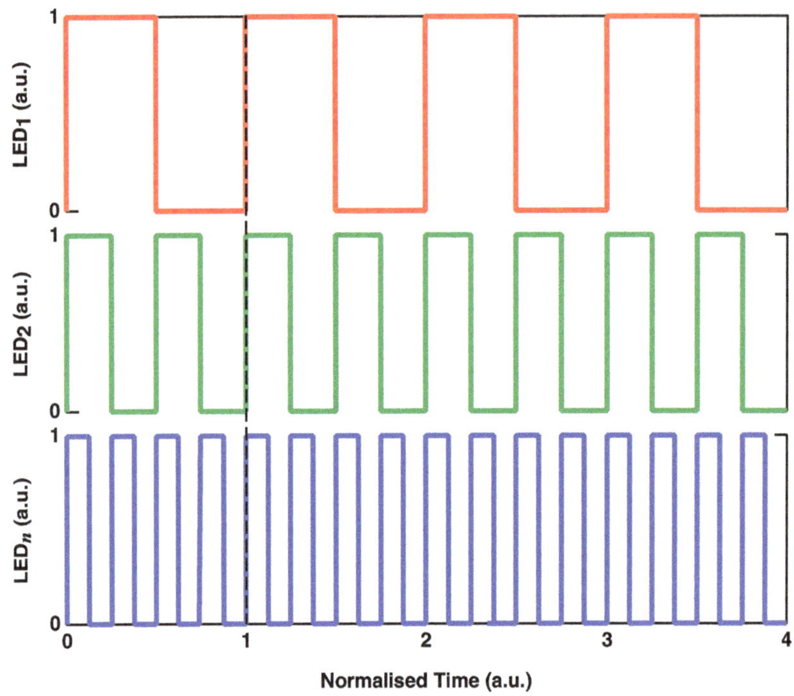

Figure 7.2. The TDMA approach to user localisation where each LED transmits a different frequency.

received power from each source can be measured, which corresponds to a specific distance based on the relationship defined in (3.20) in chapter 3. Since each LED emits with a circular footprint, the intersection of three or more LED footprints will reveal the position of the receiver. This concept is illustrated in figure 7.4, which is a three-dimensional image projected onto a two-dimensional plane. The position is revealed via a set of simultaneous Pythagoras equations in three dimensions [35]:

$$\mathbf{D} = \begin{cases} \sqrt{(x - x_1)^2 + (y - y_1)^2 + (z - z_1)^2} \\ \sqrt{(x - x_2)^2 + (y - y_2)^2 + (z - z_2)^2} \\ \sqrt{(x - x_3)^2 + (y - y_3)^2 + (z - z_3)^2} \end{cases} \tag{7.1}$$

where $[x, y, z]$ are the locations of the receiver and $[x_i, y_i, z_i]$ is the location of the ith transmitter, and $\mathbf{D} = [d_1, d_2, d_3]$ and d_i is the distance between the ith transmitter and the receiver.

It should be noted that in [35], the z part of (7.1) was dropped since it is assumed that the transmitter and receiver are on exactly the same level. However, it is well known that even a slight deviation in any of the z locations can hugely impact the accuracy of the estimated position [39, 40].

In reality, the intersections of the LED footprints rarely occur at a singular point and a more realistic trilateration scenario is described in figure 7.5, where a small

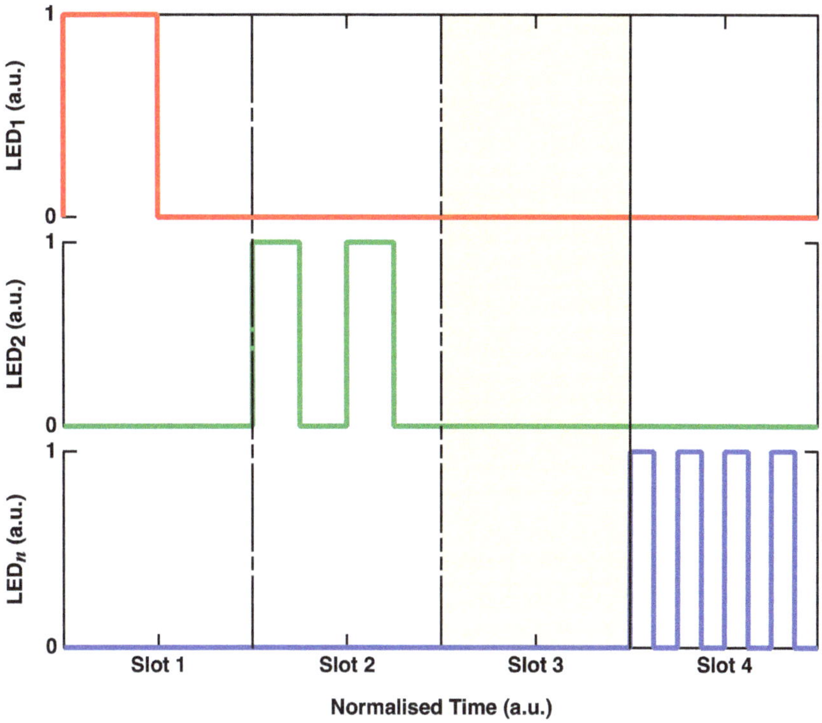

Figure 7.3. Each LED must also transmit during a different time slot.

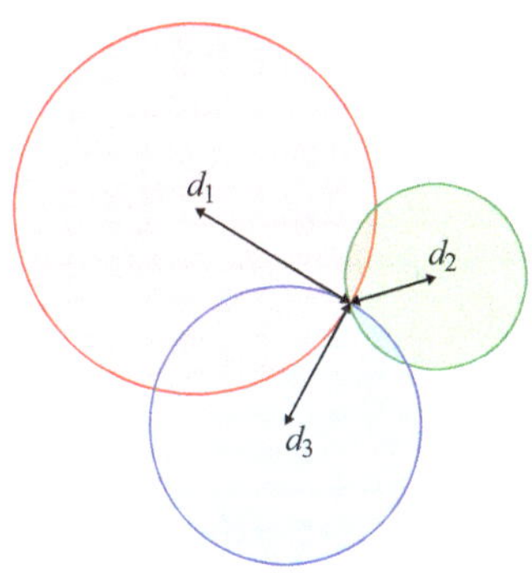

Figure 7.4. An example of perfect trilateration through the intersection of footprints.

region of interest is created inside each of the points of intersection. Adding additional sources would reduce the volume of this area, thus increasing the accuracy of the position estimation.

Fingerprinting

Fingerprinting is another method that estimates position using two steps; (i) offline survey and (ii) online positioning. During the first step, information about the environment is collected using one of the various methods discussed in this chapter,

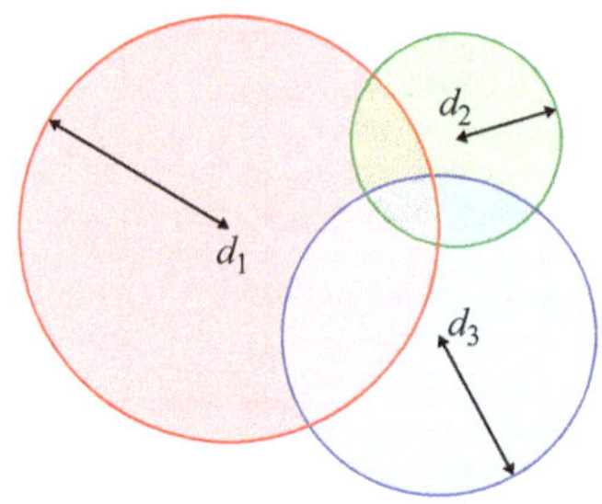

Figure 7.5. A more realistic example of trilateration that identifies a central zone through the overlapping of footprints.

either through RSS, angle of arrival (AOA) or time difference of arrival (TDOA) [21, 34, 41–43]. The latter two will be discussed later in this chapter. While this adds nothing new to the information collection, the receiver plane is divided into a matrix grid with known dimensions called fingerprint locations. In each element of the grid, the signal power is measured for each frequency and stored. This is compared via subtraction with the estimated RSS power from the theoretical relationship of the method utilised, either RSS, AOA or TDOA, as follows [40]:

$$D(\xi, r) = \sum_{f=1}^{F} \left(\xi_{m,f} - \xi_{r,f}\right)^2 \tag{7.2}$$

where D is the fingerprint database element for the rth input element, F is the number of LED frequencies, and $\xi_{m,f}$ and $\xi_{r,f}$ are the measured and reference powers, respectively. The position is estimated by searching for the grid element with the minimum value of D. There are a number of algorithms that improve the accuracy of fingerprinting, which is susceptible to variations in the optoelectronic responses of the LEDs and PDs used. These include probabilistic [44], deterministic [45, 46] or proximity-based methods [47], which aren't covered here but can be found in the appropriate references listed.

7.2.2 Angle of arrival

The next positioning scheme to be discussed is called AOA, which refers to the incident angle φ_{AOA} that the lightwave arrives at to the PD receiver, and it is discussed in [48]. This process is called triangulation (similar but not the same as trilateration). Obviously, once the incident light to the receiver has been absorbed, the information about the AOA is lost and therefore this method is not simple to implement. The AOA information is commonly obtained via image transformation, where a PD is not used as the receiver but an image sensor is instead. The image sensor is used as a camera and then, in conjunction with the known transmitter positions, uses trigonometric relationships between these locations and the relative locations of the light captured via the camera to calculate the AOA, as illustrated in figure 7.6, which shows only a single transmitter trace for simplicity. Since the dimensions of the image sensor are known, the angle of arrival can be predicted and

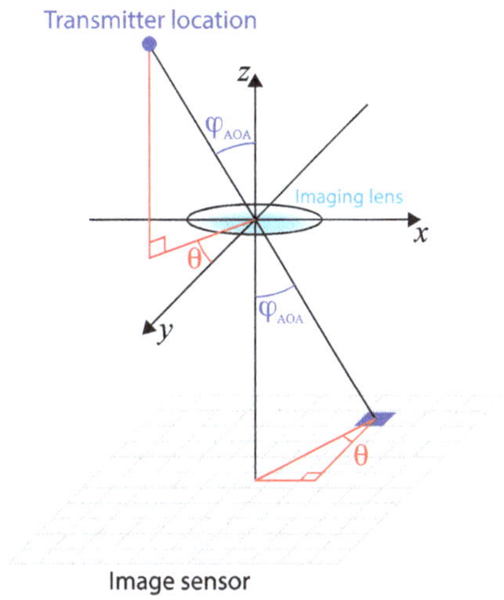

Figure 7.6. The angle-of-arrival approach that requires an image sensor with multiple elements to identify positions relative to known transmitter location. Each individual transmitter can identify itself through its transmission frequency and/or pre-defined time slot.

therefore location can be estimated accurately if there are plural transmitters with known locations.

It is also possible to obtain AOA information when using a PD alongside the channel models. In this method, since the received power is strongly correlated to the AOA via the channel models outlined in chapter 3, it is possible to calculate the receiver position if the relative position of the PD photoactive area is known. This is more difficult, however, since the required PD positioning, known locally, requires additional on-board sensors and therefore is not a very feasible option. On the other hand, AOA-based positioning systems have numerous advantages over RSS and TDOA, the main one being that they do not make use of any multiplexing and therefore synchronisation and accurate timing are less critical than in the other schemes [35].

7.2.3 Time difference of arrival

Next, TDOA will be discussed and it should be pointed out that this method can be done in two ways. The first measures the absolute time of flight from each of the transmitters to the receiver and compares the time differences and the second measures the time *difference* between the arrivals. Clearly, the former requires a significant degree of synchronisation and the timing considerations impose critical limits on the performance of the scheme [49, 50] and therefore the vast majority of the reports in the literature have employed the *difference* method for this reason [51, 52].

In general, through the nature of the method, there must be at least two PDs present in the system to be able to calculate a difference between two incoming lightwaves [35], however, researchers have overcome this using frequency-division multiplexing (FDM) and BPF [38]. As a result, TDOA can be performed using a single receiver with multiple transmitters, as always.

Using the block diagram in figure 7.7 [38], TDOA can be realised utilising trilateration to determine the position of the receiver via determination of the path differences between the transmitters and the receiver. Recalling that each LED utilises a unique orthogonal frequency, the received signal $y^i_{\mathrm{BPF}}(t)$ at the receiver from the ith transmitter after the first BPF is given by [38]:

$$y^i_{\mathrm{BPF}}(t) = \rho_i \cos\left(2\pi f_i\left[t - \frac{d_i}{c}\right] + \theta_0\right) \tag{7.3}$$

where ρ_i is a proportionality constant that is related to the channel response of the ith LED $H(0)_i$. The distance between the ith LED and the receiver is d_i, c is the speed of light and θ_0 is the initial phase delay. This phase delay is just mentioned for completeness and is not the phase difference that the TDOA scheme requires.

After the above, the signal is then down-converted to baseband and filtered again and is then given as [38]:

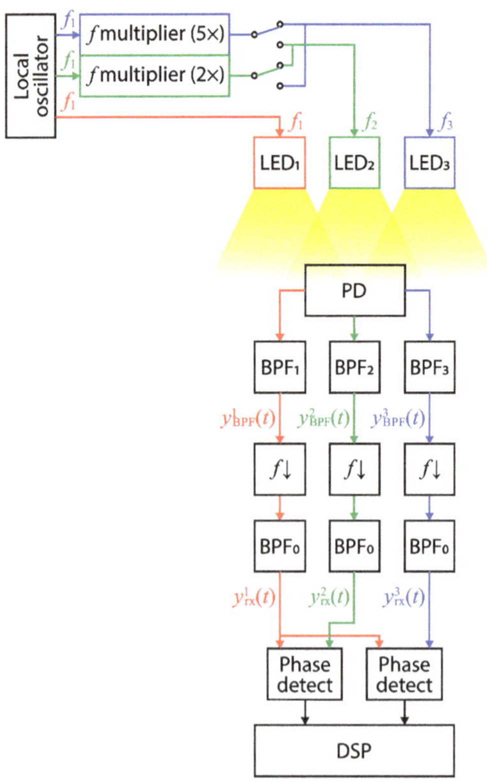

Figure 7.7. The block diagram for the time difference of the arrival system presented in [38]. Each LED is assigned its own frequency that it transmits, before band-pass filtering at the receiver frequency down-conversion and more band-pass filtering can be used to reveal the phase difference, and in turn, the receiver position.

$$y_{\rm rx}^{i}(t) = \kappa_i \cos\left(\pi\left[f_1 t - f_i \frac{d_i}{c}\right] + \theta_{\rm total}\right) \tag{7.4}$$

where κ_i is another proportionality constant. The key difference between (7.3) and (7.4), is that the 2 in the left-hand term inside the cosine drops out due to the trigonometric identity, through down-conversion to the frequency f_1. The base frequency is selected to ensure orthogonality between LEDs and typical values of $f = 1, 3, 5$ kHz, according to the literature [53].

The next step in this method is to calculate the desired phase differences via a Hilbert transform, denoted by $\mathcal{H}[.]$ and for its mathematical definition and operation, the reader is encouraged to refer to [54, 55] for more information. As noted in [35], generally only two phase differences can be obtained from the three LEDs. However, a third can be obtained via a method proposed in [38] that switches the frequencies of the first two transmitters in a predetermined time slot, which manipulates the system into providing the three desirable phase differences $\phi = [\phi_{12}, \phi_{13}, \phi_{21}]$. They are given by [38]:

$$\phi = \begin{cases} \phi_{12} = 2\pi f_1\left[d_1 - \frac{f_3}{f_1}d_2\right]c^{-1} = \tan^{-1}\left[\frac{I_{12}}{Q_{12}}\right] \\ \phi_{13} = 2\pi f_1\left[d_1 - \frac{f_5}{f_1}d_3\right]c^{-1} = \tan^{-1}\left[\frac{I_{13}}{Q_{13}}\right] \\ \phi_{21} = 2\pi f_1\left[d_2 - \frac{f_3}{f_1}d_1\right]c^{-1} = \tan^{-1}\left[\frac{I_{21}}{Q_{21}}\right] \end{cases} \tag{7.5}$$

where I_{ab} and Q_{ab} are developed in [38]:

$$I_{ab} = y_{\rm rx}^{a}(t)\mathcal{H}\left[y_{\rm rx}^{b}(t)\right] - \mathcal{H}\left[y_{\rm rx}^{a}(t)\right]y_{\rm rx}^{b}(t) \tag{7.6}$$

$$Q_{ab} = y_{\rm rx}^{a}(t)y_{\rm rx}^{b}(t) + \mathcal{H}\left[y_{\rm rx}^{a}(t)\right]\mathcal{H}\left[y_{\rm rx}^{b}(t)\right] \tag{7.7}$$

and a and b represent the transmitter of interest.

The final requirement to be able to determine positions is the distance calculation to feed into (7.5) as follows [38]:

$$\mathbf{D} = \begin{cases} d_1 = -\frac{c}{16\pi f_1}\left(\tan^{-1}\left[\frac{I_{12}}{Q_{12}}\right] + 3\tan^{-1}\left[\frac{I_{21}}{Q_{21}}\right]\right) \\ d_2 = \frac{1}{3}\left(d_1 - \frac{c}{2\pi f_1}\tan^{-1}\left[\frac{I_{12}}{Q_{12}}\right]\right) \\ d_3 = \frac{1}{5}\left(d_1 - \frac{c}{2\pi f_1}\tan^{-1}\left[\frac{I_{13}}{Q_{13}}\right]\right) \end{cases} \tag{7.8}$$

Thus, position can be estimated via trilateration using all the information provided above. There is an extension to the trilateration technique known as multilateration that is different from trilateration [56]. Instead of using multiple distance measurements as in trilateration, multilateration instead measures the differences of distance between any two LEDs whose locations are known, and this distance is given by [35]:

$$d_i - d_j = \sqrt{(x_i - x_R)^2 + (y_i - y_R)^2 + (z_i - z_R)^2} - \sqrt{(x_j - x_R)^2 + (y_j - y_R)^2 + (z_j - z_R)^2} \tag{7.9}$$

where $[x_R, y_R, z_R]$, $[x_i, y_i, z_i]$ and $[x_j, y_j, z_j]$ are the coordinates of the receiver, and the ith and jth LEDs, respectively.

Finally, we adopt table 7.1 from [35], which summarises a number of state-of-the-art VLC positioning systems and the reported details of each system. From the table it is clear that there are a wide variety of VLC-based positioning and localisation systems that can accurately position a receiver with a high degree of accuracy, at various degrees of expense. Interestingly, the accuracy does not improve with cost, as shown in table 7.1, which is sorted first by cost and then accuracy.

7.3 Positioning with data communication

The above methods have been combined with data communication on occasion [16, 58] to provide the triple functionality of room illumination, data communication *and* positioning. One such example was presented in [58] which makes use of OFDM and trilateration to transfer information and determine position simultaneously, taking advantage of the orthogonal nature of the subcarriers.

In the report [58], any three LEDs out of a wider ceiling cluster can be used to estimate the position of the receiver, as illustrated in figure 7.8. The OFDM system under test consists of $N = 128$ subcarriers modulated by 4-QAM data. The first three subcarriers are used for the trilateration and the remaining ones are used for data transmission. The test setup used is illustrated in figure 7.8 including details of the OFDM DSP setup. Since the receiver is always under at least three LEDs, it was shown that the position can always be obtained through trilateration from the three transmitters with the strongest received signal powers. The distances can be obtained through the RSS method and the distances estimated using (7.1).

A mean positioning error of 1.68 cm was found in [58], indicating a relatively high indoor localisation accuracy while providing data communications at an acceptable BER level. The location and reference estimates are illustrated in figure 7.9. They also reported that the position accuracy decreases with increasing noise power as is expected, since the RSS is more difficult to accurately determine if the noise power is of a similar absolute value.

A similar result was reported in [59], which also showed that positioning and data communication can coexist. In this report, a slightly different approach was taken to [58], as the localisation scheme was placed in between two signal bands with some guard space on either side of the positioning spectrum as illustrated in figure 7.10 for

Table 7.1. Summary of localisation systems reported in the literature.

Ref.	Method	Experimental/simulation	Testbed size (m)	Accuracy	No. devices	Noise	Cost
[32]	TDOA, trilateration	Simulation	5 × 5 × 3	0.002 cm, 3D	5 LEDs	No	Low
[38]	TDOA, trilateration	Simulation	5 × 5 × 3	0.18 cm, 2D	3 LEDs	No	Low
[57]	RSS, trilateration	Simulation	0.9 × 0.9 × 1.5	1.58 cm, 2D	3 LEDs	Yes	Low
[58]	RSS, trilateration	Experimental	0.2 × 0.2 × 0.15	1.68 cm, 3D	25 LEDs	Yes	Low
[37]	RSS, trilateration	Experimental	0.6 × 0.6 × 0.6	2.4 cm, 3D	3 LEDs	No	Low
[56]	TDOA, multilateration	Simulation	5 × 5 × 3	3.59 cm, 3D	9 LEDs	Yes	Low
[59]	RSS, fingerprinting	Experimental	1.2 × 1.2 × 2.1	~6 cm, 3D	16 LEDs	Yes	Low
[39]	AOA, triangulation	Experimental	0.71 × 0.74 × 2.26	10 cm, 3D	5 LEDs	No	Low
[33]	AOA, triangulation	Simulation	4 × 4 × 3.5	13.95 cm, 2D	16 LEDs, Inertial sensor	Yes	Low
[60]	AOA, triangulation	Simulation	4 × 6	<15 cm, 2D	4 LEDs	No	Low
[40]	RSS, trilateration	Simulation	6 × 4 × 4	<20 cm, 2D	4 LEDs	Yes	Low
[31]	RSS, trilateration	Simulation	1.5 × 1.5 × 2	<20 cm, 2D	3 LEDs	No	Low
[61]	RSS, trilateration	Experimental	5 × 8	30 cm, 2D	4 LEDs, Inertial sensor	No	Low
[62]	TDOA, multilateration	Simulation	5 × 5 × 3	31.3 cm, 3D	16 LEDs	Yes	Low
[63]	RSS, others	Simulation	10 × 2.5 × 3	<130 cm, 2D	*L* LEDs	Yes	Low
[64]	RSS, trilateration others	Experimental	2.5 × 2.84 × 2.5	0.14 m, 2D	7 LEDs, Inertial sensor	Yes	Med.
[36]	RSS, trilateration	Simulation	6 × 6 × 4.2	<15 cm, 2D	3 LEDs	Yes	Med.
[65]	AOA, trilateration	Simulation	1.8 × 1.8 × 3.5	15.6 cm, 3D	4 LEDs, Camera	No	Med.
[66]	AOA, triangulation	Experimental	5 × 3 × 3	25 cm, 3D	3 LEDs	Yes	Med.
[67]	RSS	Experimental	5 × 4 × 3	60 cm	3 LEDs, 3 PDs	Yes	Med.
[68]	RSS, filters	Simulation	‘Large’	‘High’	*L* LEDs	Yes	Med.
[69]	RSS, fingerprinting	Simulation	30 × 30	0.81 m, 2D	137 LEDs	Yes	High
[70]	RSS	Simulation	20 × 20 × 3	3 cm, 2D	9 LEDs	No	High
[43]	RSS, fingerprinting	Experimental	1.8 × 1.2 × 1	14.84 cm, 2D	6 LEDs, 4 PDs	Yes	High
[34]	RSS, fingerprinting	Simulation	6 × 6 × 4	<26.4 cm, 2D	4 LEDs	No	High

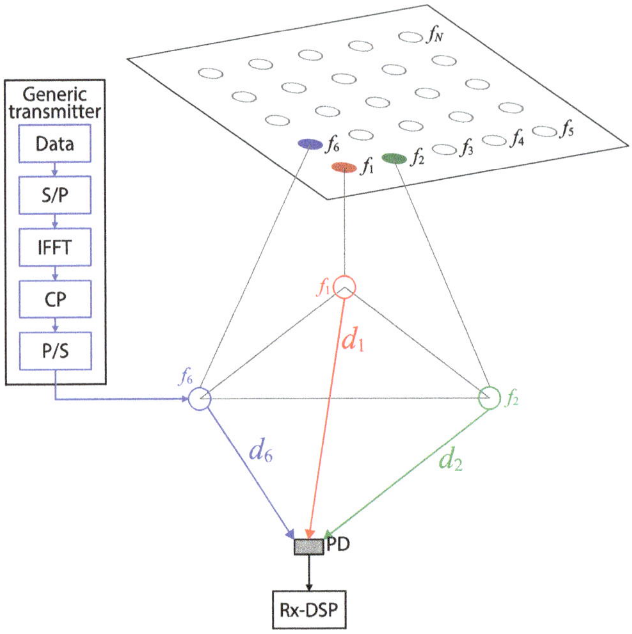

Figure 7.8. The conceptual system diagram of the system reported in [58], where any three LEDs in the ceiling can be used to localise the receiver, whilst transmitting information through other subcarriers present in the OFDM system.

both OFDM and filter-bank multi-carrier (FBMC), while the references and estimates for both schemes can be seen in figure 7.11; the mean error for the FBMC case was ~6 cm. For further information about FBMC, refer to [71].

The key result shown in [59] and the major difference from [58] is that it is shown by not using OFDM and shifting to FBMC that more of the total bandwidth available can be utilised for data communications due to the sharper roll-off of the filters used in FBMC, which are shaped using the so-called Mirabbasi–Martin filter $G(f)$, given by [59]:

$$G(f) = \sum_{k=1-K}^{K-1} \chi_k \frac{\sin(\pi(f - k/NK)NK)}{NK\sin(\pi(f - k/NK))} \tag{7.10}$$

where K is the overlapping factor, $K = 4$ in [59], χ_k is the out-of-band filter frequency response coefficient and N is the size of the IFFT used. It is shown that because of reduced power leakage in FBMC in comparison to OFDM, out-of-band interference is suppressed and the guard band spacing can be significantly reduced. This in turn means that up to 98% of available bandwidth can be used (FBMC) in comparison to 72% (OFDM), which is significant. Interestingly, the measured BER for the FBMC system was inferior to that of OFDM as can be seen in [59].

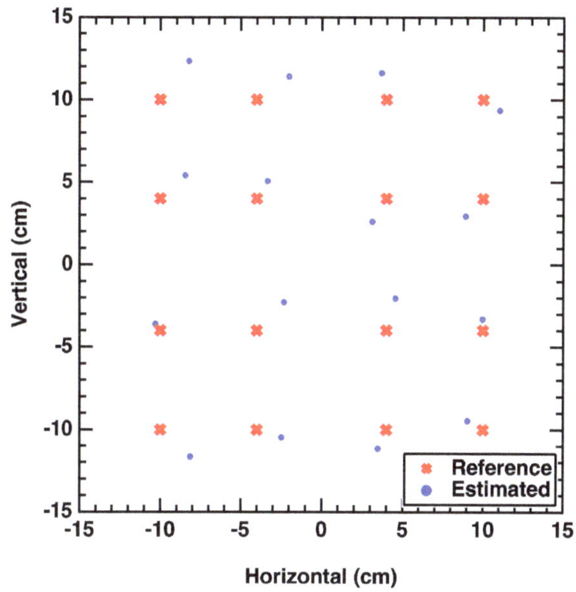

Figure 7.9. The positioning error of the report in [58], showing a high degree of accuracy in an indoor scenario.

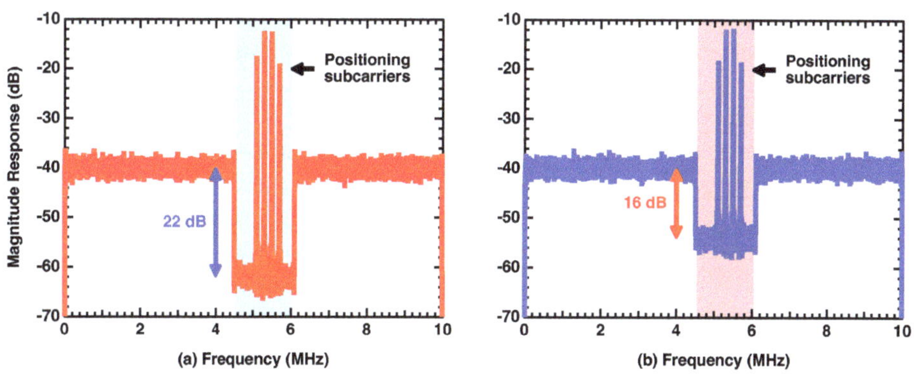

Figure 7.10. The receiver spectra of the FBMC and OFDM compatible positioning systems presented in [59]. The central frequencies are the ones used for positioning while the square envelopes are used for information transmission. As indicated in the figures, properly utilising FBMC results in improved out-of-band noise rejection and therefore a higher SNR for the localisation signals.

7.4 Summary

Chapter 7 discussed localisation and positioning systems. Starting with trilateration, the most widely used method for positioning, examples were given on how to share the channel and identify different users through time- and frequency-division multiplexing. Each user is designated a specific time slot and associated frequency. When it is the users turn to access the channel, the transmitter will transmit in their allocated resources and localisation can be achieved provided that there is at least three transmitters active at any given time.

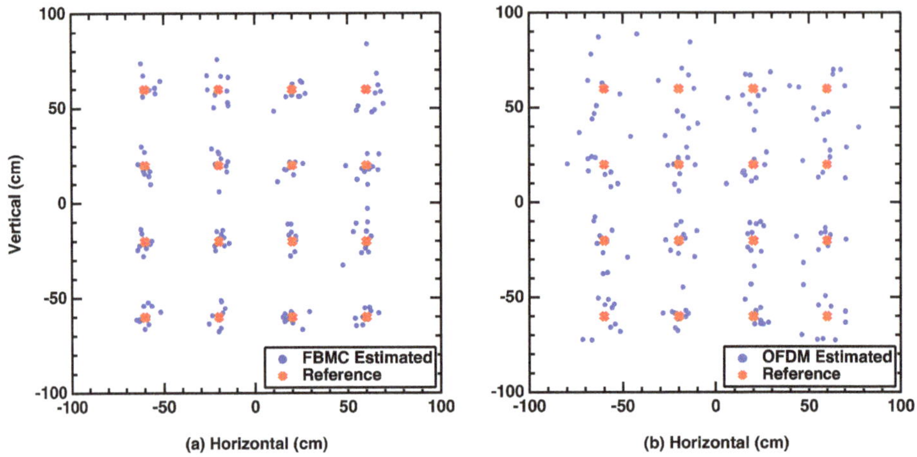

Figure 7.11. The relative positioning error for each modulation format. It is clear that the improved SNR for the FBMC subcarriers results in improved positioning accuracy.

Practical considerations to trilateration were considered, in particular that the diameters of the three LEDs' transmitters will not be perfectly overlapping at any given time, lowering precise location detection to an estimate. Nevertheless, advanced systems have been able to show centimetre level precision. In the authors opinion, localisation and positioning are one of the key technologies that will drive VLC systems in the future, since RF approaches lack accuracy indoors and localisation systems do not require huge bandwidths in the same way access networks do. Therefore, systems will remain low cost and could even use the in-built front-facing cameras that already exist in smartphones today, with no modification.

Other approaches discussed include triangulation, angle of arrival and time difference of arrival. An experimental system was presented that compared two different modulation formats experimentally that assigned several subcarriers to localisation and the remainder to data communication. It was shown that positioning error could be minimised to less than 1 m, whilst maintaining a high throughput.

References

[1] Hofmann-Wellenhof B, Lichtenegger H and Wasle E 2007 *GNSS—global navigation satellite systems: GPS, GLONASS, Galileo, and More* (Berlin: Springer)

[2] Dow J M, Neilan R E and Rizos C 2009 The international GNSS service in a changing landscape of global navigation satellite systems *J. Geod* **83** 191–8

[3] Benedicto J, Dinwiddy S, Gatti G, Lucas R and Lugert M 2000 GALILEO: Satellite system design *European Space Agency* (Citeseer) (http://citeseerx.ist.psu.edu/viewdoc/download?doi=10.1.1.200.7245&rep=rep1&type=pdf)

[4] Grajski K A and Kirk E 2003 Towards a mobile multimedia age-location-based services: a case study *Wirel. Pers. Commun.* **26** 105–16

[5] Stuart V L and Sanberg P R 2019 Queen Elizabeth prize for engineering honors three NAI fellows for innovations that changed the world *Technol. Innov.* **21** 75–84

[6] Volakis J L, O'Brien A J and Chen C-C 2016 Small and adaptive antennas and arrays for GNSS applications *Proc. IEEE* **104** 1221–32
[7] Zhang R, Biagi M, Lampe L, Little T D C, Mangold S and Xu Z 2018 Guest editorial localisation, communication and networking with VLC *IEEE J. Sel. Areas Commun.* **36** 1–7
[8] Lv H, Feng L, Yang A, Guo P, Huang H and Chen S 2017 High accuracy VLC indoor positioning system with differential detection *IEEE Photonics J.* **9** 1–13
[9] Steendam H 2017 A 3-D positioning algorithm for AOA-based VLP with an aperture-based receiver *IEEE J. Sel. Areas Commun.* **36** 23–33
[10] Zhuang Y and El-Sheimy N 2015 Tightly-coupled integration of WiFi and MEMS sensors on handheld devices for indoor pedestrian navigation *IEEE Sens. J.* **16** 224–34
[11] Lan H, Yu C, Zhuang Y, Li Y and El-Sheimy N 2015 A novel Kalman filter with state constraint approach for the integration of multiple pedestrian navigation systems *Micromachines* **6** 926–52
[12] Le Dortz N, Gain F and Zetterberg P 2012 WiFi fingerprint indoor positioning system using probability distribution comparison *2012 IEEE Int. Conf. On Acoustics, Speech and Signal Processing (ICASSP)* (IEEE) pp 2301–4
[13] Yang C and Shao H-R 2015 WiFi-based indoor positioning *IEEE Commun. Mag.* **53** 150–7
[14] Hilsenbeck S, Bobkov D, Schroth G, Huitl R and Steinbach E 2014 Graph-based data fusion of pedometer and WiFi measurements for mobile indoor positioning *Proc. 2014 ACM Int. Joint Conf. on Pervasive and Ubiquitous Computing* pp 147–58
[15] Armstrong J, Sekercioglu Y A and Neild A 2013 Visible light positioning: a roadmap for international standardization *IEEE Commun. Mag.* **51** 68–73
[16] Chen J and You X 2017 Visible light positioning and communication cooperative systems *2017 16th Int. Conf. on Optical Communications and Networks (ICOCN)* (IEEE) pp 1–3
[17] Căilean A-M and Dimian M 2017 Current challenges for visible light communications usage in vehicle applications: a survey *IEEE Commun. Surv. Tutor.* **19** 2681–703
[18] Aminikashani M, Gu W and Kavehrad M 2016 Indoor positioning with OFDM visible light communications *2016 13th IEEE Annual Consumer Communications & Networking Conf. (CCNC)* (IEEE) pp 505–10
[19] Arnon S 2015 *Visible Light Communication* (Cambridge: Cambridge University Press)
[20] Do T-H and Yoo M 2016 An in-depth survey of visible light communication based positioning systems *Sensors* **16** 678
[21] Zhuang Y, Syed Z, Li Y and El-Sheimy N 2015 Evaluation of two WiFi positioning systems based on autonomous crowdsourcing of handheld devices for indoor navigation *IEEE Trans. Mob. Comput.* **15** 1982–95
[22] Zhuang Y, Yang J, Li Y, Qi L and El-Sheimy N 2016 Smartphone-based indoor localization with bluetooth low energy beacons *Sensors* **16** 596
[23] Hossain A K M M and Soh W-S 2007 A comprehensive study of bluetooth signal parameters for localization *2007 IEEE 18th Int. Symp. on Personal, Indoor and Mobile Radio Communications* (IEEE) pp 1–5
[24] IEEE Standard Association *et al* 2011 IEEE standard for local and metropolitan area networks-part 15.7: short-range wireless optical communication using visible light (Piscataway, NJ: IEEE) p 1–309
[25] Tsonev D, Videv S and Haas H 2014 Light fidelity (Li-Fi): towards all-optical networking *Proc. SPIE* **9007** 900702

[26] Karunatilaka D, Zafar F, Kalavally V and Parthiban R 2015 LED based indoor visible light communications: state of the art *IEEE Commun. Surv. Tutor.* **17** 1649–78
[27] Bian R, Tavakkolnia I and Haas H 2019 15.73 Gb/s visible light communication with off-the-shelf LEDs *J. Lightwave Technol.* **37** 2418–24
[28] Magno M, Polonelli T, Benini L and Popovici E 2014 A low cost, highly scalable wireless sensor network solution to achieve smart LED light control for green buildings *IEEE Sens. J.* **15** 2963–73
[29] Fan J, Yung K-C and Pecht M 2012 Lifetime estimation of high-power white LED using degradation-data-driven method *IEEE Trans. Device Mater. Reliab.* **12** 470–7
[30] Feng L, Hu R Q, Wang J, Xu P and Qian Y 2016 Applying VLC in 5G networks: architectures and key technologies *IEEE Netw.* **30** 77–83
[31] Yang S-H, Jeong E-M, Kim D-R, Kim H-S, Son Y-H and Han S-K 2013 Indoor three-dimensional location estimation based on LED visible light communication *Electron. Lett.* **49** 54–6
[32] Nadeem U, Hassan N U, Pasha M A and Yuen C 2014 Highly accurate 3D wireless indoor positioning system using white LED lights *Electron. Lett.* **50** 828–30
[33] Prince G B and Little T D C 2012 A two phase hybrid RSS/AoA algorithm for indoor device localization using visible light *2012 IEEE Global Communications Conf. (GLOBECOM)* (IEEE) pp 3347–52
[34] Nadeem U, Hassan N U, Pasha M A and Yuen C 2015 Indoor positioning system designs using visible LED lights: performance comparison of TDM and FDM protocols *Electron. Lett.* **51** 72–4
[35] Zhuang Y, Hua L, Qi L, Yang J, Cao P, Cao Y, Wu Y, Thompson J and Haas H 2018 A survey of positioning systems using visible LED lights *IEEE Commun. Surv. Tutor.* **20** 1963–88
[36] Gu W, Zhang W, Kavehrad M and Feng L 2014 Three-dimensional light positioning algorithm with filtering techniques for indoor environments *Opt. Eng.* **53** 107107
[37] Kim H-S, Kim D-R, Yang S-H, Son Y-H and Han S-K 2012 An indoor visible light communication positioning system using a RF carrier allocation technique *J. Lightwave Technol.* **31** 134–44
[38] Jung S-Y, Hann S and Park C-S 2011 TDOA-based optical wireless indoor localization using LED ceiling lamps *IEEE Trans. Consum. Electron.* **57** 1592–7
[39] Kuo Y-S, Pannuto P, Hsiao K-J and Dutta P 2014 Luxapose: indoor positioning with mobile phones and visible light *Proc. 20th Annual Int. Conf. on Mobile Computing and Networking* pp 447–58
[40] Zhang W, Chowdhury M I S and Kavehrad M 2014 Asynchronous indoor positioning system based on visible light communications *Opt. Eng.* **53** 045105
[41] Kushki A, Plataniotis K N, Venetsanopoulos A N and Regazzoni C S 2005 Radio map fusion for indoor positioning in wireless local area networks *2005 7th Int. Conf. on Information Fusion* vol 2 (IEEE) pp 8
[42] Youssef M and Agrawala A 2008 The Horus location determination system *Wireless Netw.* **14** 357–74
[43] Vongkulbhisal J, Chantaramolee B and Zhao Y 2012 A fingerprinting-based indoor localization system using intensity modulation of light emitting diodes *Microw. Opt. Technol. Lett.* **54** 1218–27
[44] Tardos G 2008 Optimal probabilistic fingerprint codes *J. ACM (JACM)* **55** 1–24

[45] Dawes B and Chin K-W 2011 A comparison of deterministic and probabilistic methods for indoor localization *J. Syst. Softw.* **84** 442–51
[46] Honkavirta V, Perala T, Ali-Loytty S and Piché R 2009 A comparative survey of WLAN location fingerprinting methods *2009 6th Workshop on Positioning, Navigation and Communication* (IEEE) pp 243–51
[47] Faragher R and Harle R 2015 Location fingerprinting with bluetooth low energy beacons *IEEE J. Sel. Areas Commun.* **33** 2418–28
[48] Li M and Lu Y 2008 Angle-of-arrival estimation for localization and communication in wireless networks *2008 16th European Signal Processing Conf.* (IEEE) pp 1–5
[49] Huang Q, Ghogho M, Wei J and Ciblat P 2010 Practical timing and frequency synchronization for OFDM-based cooperative systems *IEEE Trans. Signal Process.* **58** 3706–16
[50] Ghebretensaé Z, Harmatos J and Gustafsson K 2010 Mobile broadband backhaul network migration from TDM to carrier Ethernet *IEEE Commun. Mag.* **48** 102–9
[51] Shen G, Zetik R and Thoma R S 2008 Performance comparison of TOA and TDOA based location estimation algorithms in LOS environment *2008 5th Workshop on Positioning, Navigation and Communication* (IEEE) pp 71–8
[52] Kaune R 2012 Accuracy studies for TDOA and TOA localization *2012 15th Int. Conf. on Information Fusion* (IEEE) pp 408–15
[53] Armstrong J 2009 OFDM for optical communications *J. Lightwave Technol.* **27** 189–204
[54] Cizek V 1970 Discrete Hilbert transform *IEEE Trans. Audio Electroacoust.* **18** 340–3
[55] Selesnick I W 2001 Hilbert transform pairs of wavelet bases *IEEE Signal Process. Lett.* **8** 170–3
[56] Do T-H, Hwang J and Yoo M 2013 TDoA based indoor visible light positioning systems *2013 Fifth Int. Conf. on Ubiquitous and Future Networks (ICUFN)* (IEEE) pp 456–8
[57] Yang S-H, Kim D-R, Kim H-S, Son Y-H and Han S-K 2013 Visible light based high accuracy indoor localization using the extinction ratio distributions of light signals *Microw. Opt. Technol. Lett.* **55** 1385–9
[58] Lin B, Tang X, Ghassemlooy Z, Lin C and Li Y 2017 Experimental demonstration of an indoor VLC positioning system based on OFDMA *IEEE Photonics J.* **9** 1–9
[59] Yang H, Chen C, Zhong W-D, Alphones A, Zhang S and Du P 2018 Demonstration of a quasi-gapless integrated visible light communication and positioning system *IEEE Photonics Technol. Lett.* **30** 2001–4
[60] Eroglu Y S, Guvenc I, Pala N and Yuksel M 2015 Aoa-based localization and tracking in multi-element VLC systems *2015 IEEE 16th Annual Wireless and Microwave Technology Conf. (WAMICON)* (IEEE) pp 1–5
[61] Li L, Hu P, Peng C, Shen G and Zhao F 2014 Epsilon: a visible light based positioning system *11th USENIX Symp. on Networked Systems Design and Implementation (NSDI '14')* pp 331–43
[62] Taparugssanagorn A, Siwamogsatham S and Pomalaza-Ráez C 2013 A hexagonal coverage LED-ID indoor positioning based on TDOA with extended Kalman filter *2013 IEEE 37th Annual Computer Software and Applications Conf.* (IEEE) pp 742–7
[63] Lou P, Zhang H, Zhang X, Yao M and Xu Z 2012 Fundamental analysis for indoor visible light positioning system *2012 1st IEEE Int. Conf. on Communications in China Workshops (ICCC)* (IEEE) pp 59–63
[64] Li Z, Yang A, Lv H, Feng L and Song W 2017 Fusion of visible light indoor positioning and inertial navigation based on particle filter *IEEE Photonics J.* **9** 1–13

[65] Rahman M S, Haque M M and Kim K-D 2011 Indoor positioning by LED visible light communication and image sensors *Int. J. Electr. Comput. Eng.* **1** 161

[66] Yasir M, Ho S-W and Vellambi B N 2014 Indoor positioning system using visible light and accelerometer *J. Lightwave Technol.* **32** 3306–16

[67] Yasir M, Ho S-W and Vellambi B N 2015 Indoor position tracking using multiple optical receivers *J. Lightwave Technol.* **34** 1166–76

[68] Zheng D, Cui K, Bai B, Chen G and Farrell J A 2011 Indoor localization based on LEDs *2011 IEEE Int. Conf. on Control Applications (CCA)* (IEEE) pp 573–8

[69] Kail G, Maechler P, Preyss N and Burg A 2014 Robust asynchronous indoor localization using LED lighting *2014 IEEE Int. Conf. on Acoustics, Speech and Signal Processing (ICASSP)* (IEEE) pp 1866–70

[70] Hou Y, Xiao S, Zheng H and Hu W 2015 Multiple access scheme based on block encoding time division multiplexing in an indoor positioning system using visible light *J. Opt. Commun. Netw.* **7** 489–95

[71] Bellanger M *et al* 2010 FBMC physical layer: a primer *Phydyas* **25** 7–10

Part II

Further applications beyond communications

Chapter 8

Internet-of-things and intelligent transport systems

8.1 Introduction

One example that makes effective VLC systems is IoT, which shows huge advantages in comparison to RF systems. In general, the term 'IoT systems' refers to the interaction between an arrangement of sensors, robots or automated systems in an indoor or outdoor environment [1]. The IoT paradigm considers the ubiquitous interconnection of everything, and encompasses several other areas including machine-to-machine (M2M) [2, 3] and intelligent transport system (ITS) [4], which are specific subsets of the technology. The former refers exclusively to an industrial environment where large machines are provided with instructions to execute and the latter concerns the interconnection of vehicles to any other infrastructure.

There have been applications of IoT in VLC systems across numerous domains, including but not limited to: (i) smart homes, where bursty sensor data can be transferred over visible light to a ceiling-based router to automate home appliances, etc [5, 6]; (ii) healthcare, where in-home monitoring such as fall detection can be performed using advanced positioning algorithms [7, 8] (see chapter 7), or through/under-skin healthcare monitoring [9, 10] (see chapter 9); (iii) ITS, where safety sensors and communication can be integrated into the head and tail lights on moving vehicles [11] and finally; (iv) industrial machine communications, where large robotics can be controlled without the requirement for human intervention, thus improving safety [12]. This chapter will mainly provide insight into the ITS element. Before that, however, a general overview of IoT in VLC is provided and references are provided for further reading.

8.2 Internet-of-things

The general conceptual block diagram for IoT approximates the illustration in figure 8.1 and either focuses on a high degree of interconnectivity between all devices

doi:10.1088/978-0-7503-1680-4ch8

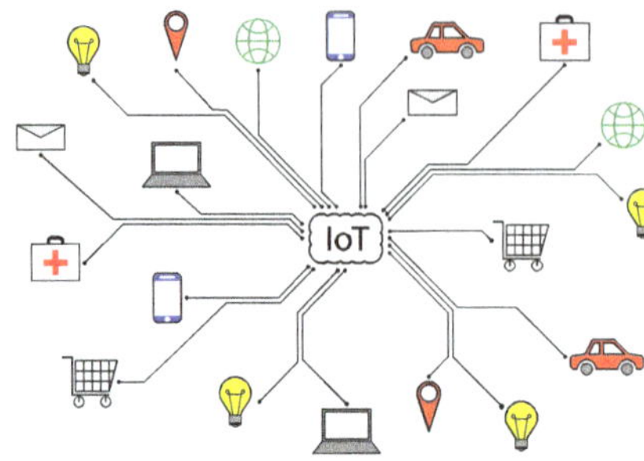

Figure 8.1. A general conceptual diagram of IoT systems.

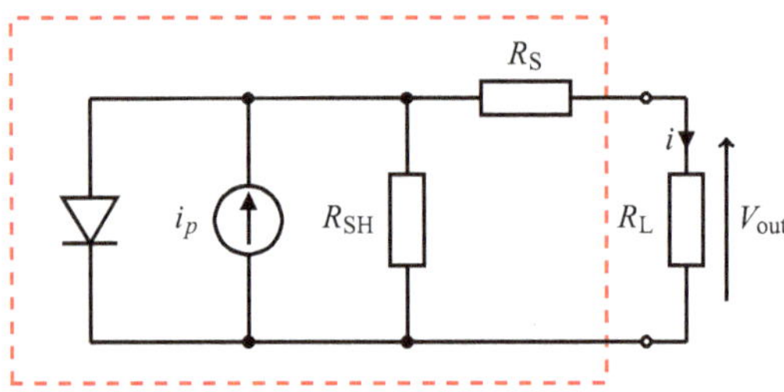

Figure 8.2. The simplified equivalent circuit of a PD considering energy harvesting.

to a base-station node, or on ad-hoc networks established device-to-device [13, 14]. No data communication aspects of IoT will be covered in this chapter since they use the fundamental aspects covered in the others.

8.2.1 Energy harvesting

Furthermore, energy harvesting for self-powered IoT applications has become an area of increasing interest, since in general the devices used are low-power and can utilise gated electronics to exist in a 'hybrid-on' state where they wake up periodically to send information whilst otherwise remaining in the 'off' state [15]. Energy harvesting in VLC was a pertinent proposition because of the dc-biasing that is introduced into the system to satisfy the non-negative optical power constraint, which is otherwise wasted energy [16]. The first report into using a photovoltaic (PV) (solar) cell was proposed in [17] where an OPV was biased and used as an OPD to detect signals, but no energy harvesting was considered and the dc was discarded. The first complete report that utilised a PV in a VLC system explicitly for energy harvesting was reported in [18] using a Si-PV and in [19] using an OPV.

The equivalent circuit for PV cells is split into two parts; the first for energy harvesting and the second for communications and they both have some resemblance to those shown for PDs in chapter 2. In figures 8.2 and 8.3 the equivalent circuits for energy harvesting and communications are drawn, respectively, where the red dashed box is indicative of the relevant equivalent circuit.

For the energy harvesting side, the conventional dc models are adopted [18, 20, 21]. For the communication system, the frequency response must be considered and

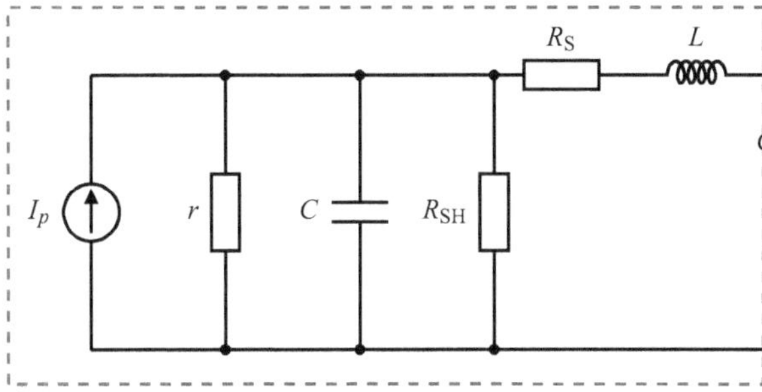

Figure 8.3. The simplified equivalent circuit of a PD considering communication systems.

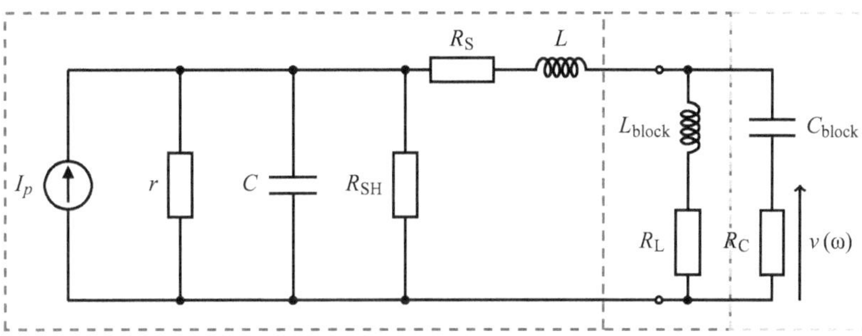

Figure 8.4. The equivalent circuit when considering the PD (red dashed box), energy harvesting arm (blue dashed box) and communications harvesting arm (green dashed box).

hence the key difference in the circuit diagram is that a parallel capacitor C is added next to the shunt resistance R_{SH}, an inductor L is added in series to the contact resistance R_S, while the series capacitor C_{block} is added to block the dc. The frequency response can be found by [18]:

$$\left|\frac{V_{out}(\omega)}{i_p(\omega)}\right|^2 = \left|\frac{R_L[R_S + j\omega L + X_{C_{block}} + R_L]^{-1}}{r^{-1} + X_C^{-1} + R_{SH}^{-1} + [R_S + j\omega L + X_{C_{block}} + R_L]^{-1}}\right|^2 \tag{8.1}$$

where $V_{out}(\omega)$ is the output ac voltage, $i_p(\omega)$ is the photocurrent generated by the incident light, $X_C = (j\omega C)^{-1}$ is the reactance of a capacitor, $\omega = 2\pi f$ is the angular frequency and R_L is the load resistor. For information on the method to select the value of these components, please refer to [18].

The overall circuit diagram for the solar panel that considers both ac and dc extraction is illustrated in figure 8.4 with each arm highlighted, where the blue and green dashed boxes are the dc and ac recovery arms, respectively. Following the circuit diagram, the overall frequency response is given by [18]:

$$\left|\frac{v(\omega)}{i_p(\omega)}\right|^2 = \left|\frac{R_C R_{LC}[(R_S + j\omega L + R_{LC})(X_{C_{block}} + R_C)]^{-1}}{r^{-1} + X_C^{-1} + R_{SH}^{-1} + [R_S + j\omega L + R_{LC}]^{-1}}\right|^2 \tag{8.2}$$

where the new term R_{LC} has been introduced, and is the resistance of the impedance network following the series inductor L, given by [18]:

$$R_{\mathrm{LC}} = \left[(j\omega L_{\mathrm{block}} + R_{\mathrm{L}})^{-1} + \left(X_{C_{\mathrm{block}}} + R_{\mathrm{C}} \right)^{-1} \right]^{-1} \tag{8.3}$$

A full noise analysis and noise-equivalent circuits are developed in [18] but are not covered here. The measured and fitted magnitude responses of the PV cell used in [18] are illustrated in figure 8.5 for different load resistances R_{L} and blocking capacitors of (a) 1 μF and (b) 100 μF, respectively. Clearly, as the load resistance decreases, the frequency response increases at the cost of overall gain, as is expected

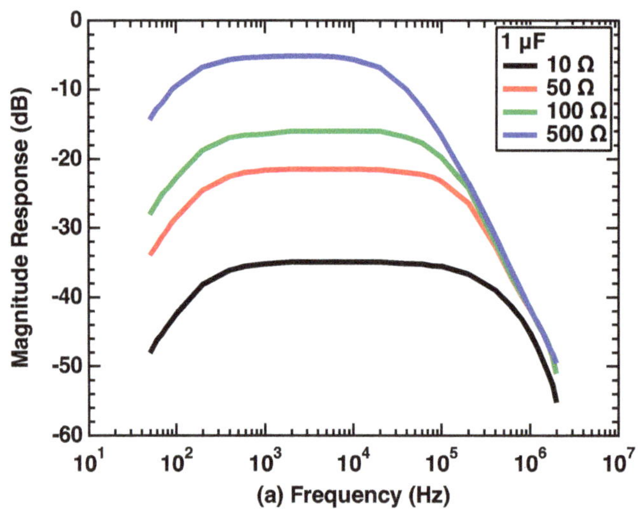

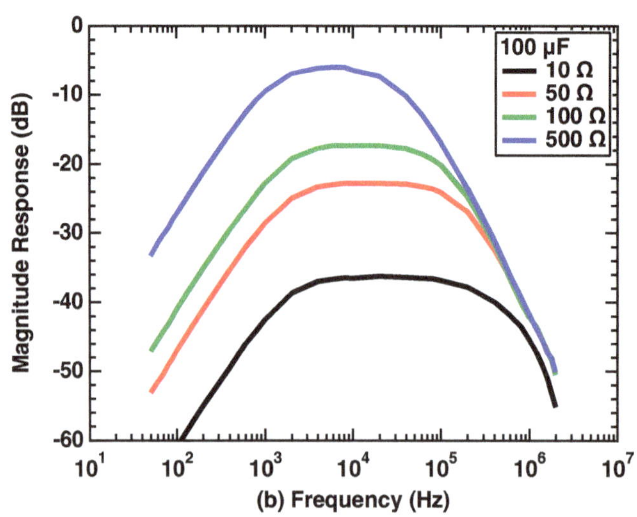

Figure 8.5. The magnitude responses of the energy harvesting system reported in [18], for (a) 1 μF and (b) 100 μF blocking capacitors. The resistances shown are the load resistors.

according to the literature [22]. At a maximum, a frequency response in the region of 1 MHz can be inferred. Similar results with an OPV device are recorded in [19].

A study in [23] reports on the levels of energy that can be harvested using the above approach using various different types of optics, including a reflector, lens, parabolic mirror placed in front of the solar panel, and these are shown as a function of distance in figure 8.6 [23]. It is immediately clear that the parabolic mirror design yields the highest energy recovery for the link since it collects the highest amount of light, and can recover approximately 20–30 dBm of power up to 5 m distance. Photographs of each of the designed optics are shown in [23].

Other reports have gone one step further and utilise the energy recovered to self-bias the receiver circuitry, whilst supporting relatively high data rates of around 20 Mb/s [24]. In [25, 26], an investigation was performed that derives relationships to jointly select the dc bias and power in order to optimise the information and energy transfer that can be obtained.

8.3 Intelligent transport systems

Vehicle applications are perhaps one of the most likely success stories of future VLC networks. As automobile technology moves forwards, there is significant progress in the field of driver-less cars [27–30] and sensors for safety applications, using both RF and VLC technologies. As European countries move to change laws to have car headlights ‘always-on’, there is an inherent use-case for VLC systems for car-to-car and car-to-infrastructure communication in order to ensure safety and road-level mapping [31–33]. Such safety interventions are desperately needed as road traffic accidents are one of the leading causes of death for the general population and is indeed the leading cause of death for people aged 15–29 [34–36]. Therefore, improving safety in vehicles is obviously paramount and special emphasis is being placed in this area. Car manufacturers constantly look at methods to improve their

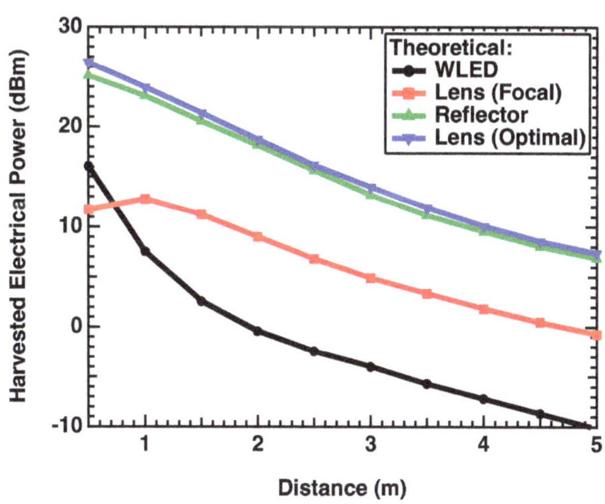

Figure 8.6. The ideal recovered harvested energy for different methods of energy harvesting, including lenses, reflectors and parabolic mirrors reported in [23].

cars mechanical safety, and the 21st century has brought a revolution in accident prevention via adding additional sensors to vehicles, introducing prescient safety considerations to cars, and the statistics report that up to 81% of accidents could be avoided if the relevant sensors were attached [37].

The ITS paradigm can be generically divided into three different subsections: (i) vehicle-to-vehicle (V2V); (ii) vehicle-to-infrastructure (V2I) and (iii) vehicle-to-X (V2X), where the final one concerns any vehicle-to-other links. The general block diagram of an ITS system is slightly more complex than the conventional VLC equivalent and it is shown in figure 8.7 [32]. The additional blocks are an automatic gain control (ACG) amplifier with an in-built limiting amplifier stage and also the inclusion of atmospheric sensor data, which is either used to steer the receiver FoV or to estimate atmospheric effects, which are of general concern in outdoor optical wireless links [38–41]. The ACG is introduced because the transmission distances are constantly varying as vehicles move closer and away from the source and hence the signal amplitude levels also vary. The ACG and limiting amplifier combination ensure that the same voltage levels are recovered regardless of the distance, but without any guarantee of an equivalent SNR.

8.3.1 Distances and mobility

One of the major challenges in using RF for safety considerations in vehicles is the nature of the frequency and channel, since radio signals pass through objects, the signals recovered are often heavily impaired by multi-path effects, which adds a challenge to accurately recovering any useful information.

This is contrary to VLC, where the lightwaves do not pass through objects and can be highly directional, avoiding NLOS reflections. In figure 8.8, a number of potential VLC sources are shown and a typical usage scenario is shown in figure 8.9. In this paradigm, the LED brake lights on car ‘A’ contain information that is collected by a photodetector on the bumper of car ‘B’ that overrides the manual controls of the second car, causing it to gently brake, avoiding a potentially fatal crash, rather than relying on the awareness of the driver. Other information can be

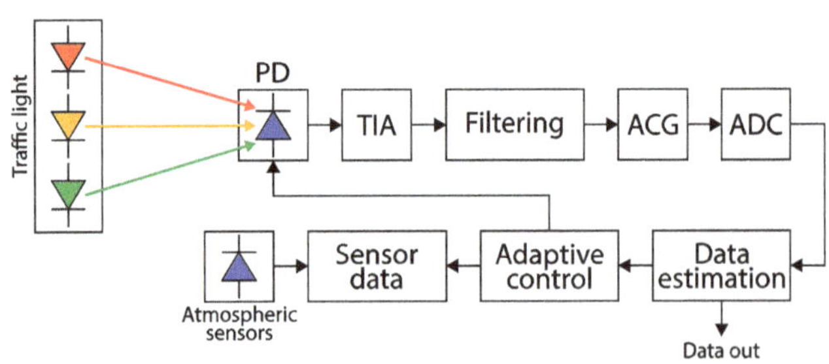

Figure 8.7. The block diagram for an infrastructure-to-vehicle ITS system. Additional components are included in comparison to indoor broadcasting systems which are the ACG, that ensures a constant voltage amplitude is recovered, at the cost of SNR, and adaptive control that reacts to the ambient sensor data and adjusts the control parameters of the receiver.

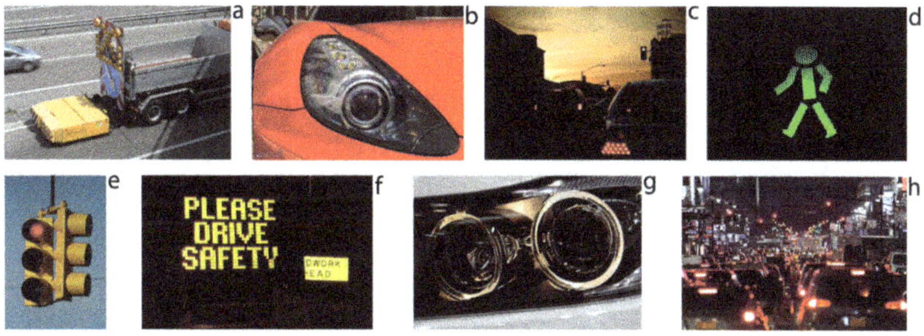

Figure 8.8. Examples of ITS transmitters. Copyright: (a) 'Roadworks—Collision Protection' by Kecko is licensed with CC BY 2.0; (b) 'Ferrari California Headlight' by [Mixtography] is licensed with CC BY-ND 2.0; (c) 'Bakersfield Brake Light' by uncleboatshoes is licensed with CC BY-ND 2.0; (d) 'Green traffic light' by Hugo-photography is licensed with CC BY-ND 2.0; (e) 'Traffic lights, Grand Rapids' by WayShare is licensed with CC BY-ND 2.0; (f) 'Road Sign Grammar' by Damien Ayers is licensed with CC BY 2.0; (g) 'The Famous BMW headlights' by prosto photos is licensed with CC BY-ND 2.0; (h) 'Traffic. Lights.' by Davide Gabino (aka Strolic Furlan) is licensed with CC BY-ND 2.0. To view a copy of these licenses, visit https://creativecommons.org/licenses/.

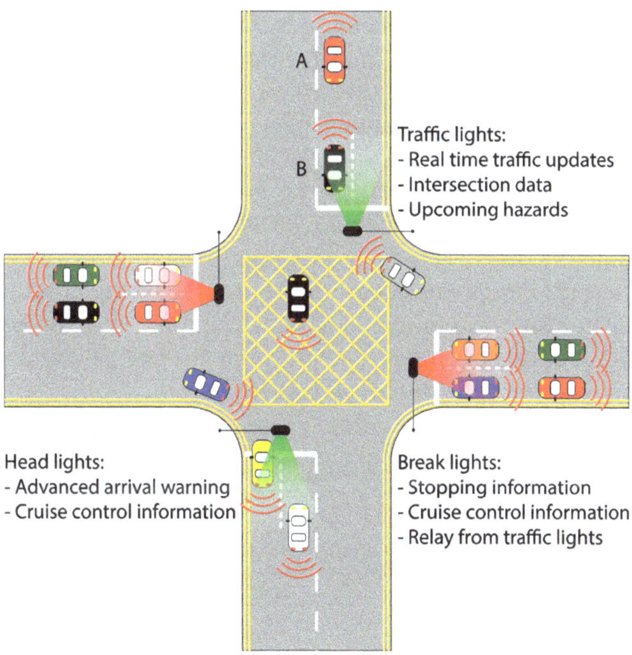

Figure 8.9. A typical ITS usage scenario, where traffic lights provide real time traffic updates, intersection data, upcoming hazards, etc. Vehicle headlights can provide advanced warning of their arrival behind other cars and cruise control information. and finally, break lights can provide early warning of breaking and can relay information from traffic lights down a queue of cars.

gathered from the traffic lights or lamp posts including mapping data and up-to-date traffic information [42].

Furthermore, one of the major issues with ITS applications is the range of up to 1000 m that must be supported, along with a number of other requirements as defined by the Vehicle Safety Communications Consortium. However, realistically, requirements for distances up to 1000 m are exaggerated since the odds of two vehicles spaced 1 km apart having an accident are extremely low. Nonetheless, these definitions led to the IEEE 802.11p standard, which was developed for radio systems operating at 5.9 GHz [32, 43–45]. More realistically are the distances reported in table 8.1, which show that a maximum distance of approximately 160 m can realistically be considered [43].

This is still a relatively high distance and it is specifically to stress test RF systems due to the high degree of interference noise that will be generally accrued over a long multi-path channel introducing reliability concerns [46–49], but VLC can largely overcome that by integrating high powered, highly directional LEDs or LDs into the lights [11, 50–52] and/or using different types of receivers such as image sensors [53]. A number of reports have been collated in table 8.2. Interestingly, a high number of reports have utilised Manchester coding in their systems, which is not a common option in general VLC systems, however, it offers simple clock recovery and synchronisation properties due to the guaranteed transition between bits and hence is a popular option for VLC-ITS systems. Clearly there are numerous reports in the literature that show high distance transmission, with particular emphasis on the work done in Japan at *Shizuoka University* in collaboration with Toyota [54–56], who have achieved transmission distances greater than 100 m, albeit at very low data rates, which is perceivably a disadvantage. However, by reducing rates, image sensors with larger photoactive areas can be used, since their bandwidths are less important and hence, higher distances can be achieved.

There is a trade-off between receiver area and likelihood of PD saturation [62], which leads to several design considerations that must be taken into account when developing receivers for this context. The most popular way to control this is to introduce a limiting amplifier circuit to control the receiver gain and ameliorate this problem and introduce sufficiently low BERs [59].

Table 8.1. Typical usage scenarios including that define approximate distances between vehicles.

Condition	Inter-vehicle distance (m)
Traffic jam	<35
Urban road	35–49
Urban highway	50–66
Rural highway	67–100
Rural road	>160

Table 8.2. A number of ITS systems reported in the literature, demonstrating either low rate systems with high reliability over long distances or high speeds over short distances, generally.

Ref.	Receiver	Receiver FoV (°)	Data rate	Distance (m)	Modulation/ coding
[56]	Image sensor	22	10, 15, 20 Mb/s	1	Manchester
[55]	Image sensor	22	10 Mb/s	7.781	Manchester
[54]	Image sensor	22	32 kb/s	70	Manchester
[54]	Image sensor	22	2 kb/s	110	Manchester
[57]	Image sensor	22	55 Mb/s	1.5	OFDM
[58]	Image sensor		375 Mb/s	2.6	OFDM
[59]	PD	10	15 kb/s	50	Manchester
[60]	PD and image sensor	1.3	4.8 kb/s	90	2-PPM
[61]	PD and image sensor	0.4	2 Mb/s	60	16-QAM
[62]	PD	20	15 kb/s	15	Manchester
[63–65]	PD		20 kb/s	50	Direct sequence spread spectrum
[66]	PD	10	15 kb/s	23	Manchester
[67]	PD		115 kb/s	31	OOK
[68]	PD		2.4 Gb/s	1.1	CAP

Another challenge faced by ITS systems is mobility, since the environment is constantly changing and passing vehicles may continually alter the communication channel. One of the first reports to address this challenge was published in [61] in 2009, where a tracking mechanism is integrated into the overall system to control the positional relationship between the transmitter and receiver. The system proposed utilised the block diagram illustrated in figure 8.10, where a wide-angled camera is used to find the broad position of the transmitter, and two mirror galvanometers are placed to deflect the incoming signal beam and extract the vertical and horizontal position information.

The beam is then focused using a lens and split, with half the power collected by a PD and half by an image sensor. The information provided by the image sensor in conjunction with the galvanometers allow high-speed direction control since it is possible to understand the location of the source of the beam from their relative outputs. The PD collects the information that was transmitted. Using this advanced tracking system, a car-to-car communication link was able to be supported up to 1 Mb/s up to 100 m.

A simpler (and cheaper) method of doing this would be to attach a PD on each side of the vehicle, as was done in [69], which guarantees that at least one receiver will be pointing in the direction of the transmitter at any given time. Interestingly, however, there is a trade-off between FoV and distance that is related to relationships found in (3.20) and (3.21), which show that the channel response is inversely

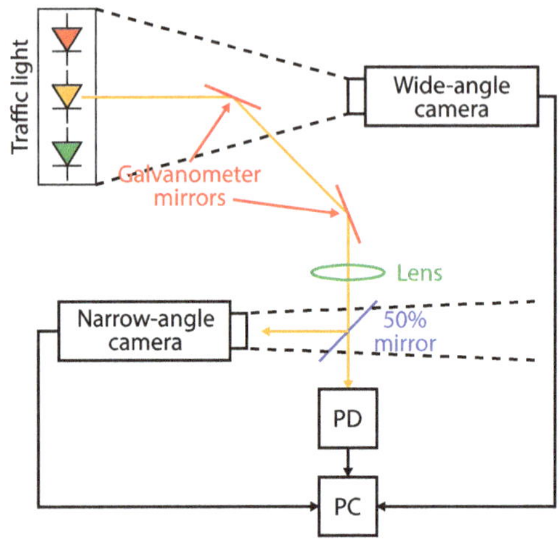

Figure 8.10. The block diagram of an optical system capable of tracking the positional relationship between the transmitter and receiver. The receiver uses two galvanometers to deflect the beam and feedback information about the incoming optical beam to the receiver. Simultaneously, the incident light beam is partially reflected through a 50% mirror and recovered using a PD and also fed into a narrow-angle camera. The information from the incident light beam plus the galvanometer mirrors can be combined to track the relative position through a wide-angled camera.

proportional to the square of the distance and also falls away with the cosine of the FoV. Therefore, to increase distances to realistic levels, a high number of PDs must be added with narrow FoV, which in turn increases the overall cost.

A solution to this problem was proposed in [66], where a cooperative scheme, adopted based on information relay, is presented. If a traffic light is updating vehicles in a queue with traffic information, it clearly cannot transmit the same signal to all the cars present because the LOS will be naturally blocked for some more than others, and each one will have a varying distance and SNR. As such, the system presented in [66] demonstrates a cooperative ITS system where the traffic light is only aiming to transmit information to the first car in the queue and that car transmits through its tail lights to the one behind, etc., until the end of the queue has obtained the information. This concept is illustrated in figure 8.11.

The report was an experimental demonstration based on a static traffic light and a simulated car transceiver. The modulation format selected was OOK with Manchester encoding transmitted with a clock frequency of 15 kHz. The data rate was particularly low because it was supported by a low-cost microcontroller that transmits/recovers the information at each node and a higher data rate was not necessary. The data is divided into a frame structure that is outlined in figure 8.12, which is the structure adopted by numerous ITS systems [66]. The traffic light was separated from the first car by a distance of 20 m. The car behind it was initially spaced at a distance of 1 m and no errors were found. In fact, an acceptable BER was found at every distance tested, with a maximum of 10^{-7} being recorded up to 23 m in [66] and 50 m in [70], which can be considered error free.

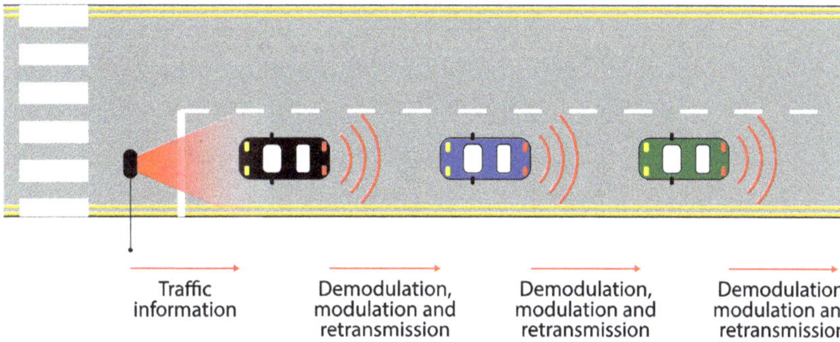

Figure 8.11. An example of platooning where the SNR may be insufficient for cars at the back of the platoon to recover the information from the traffic light. The cars sequentially receive the information from the infrastructure/vehicle in front, demodulate it, then re-transmit to the vehicle behind.

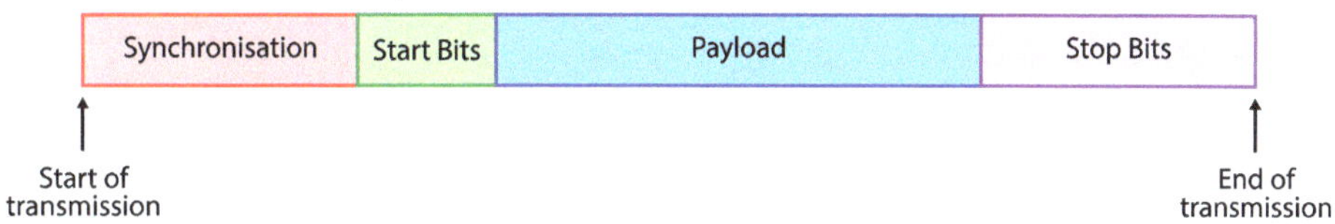

Figure 8.12. The packet framework that ITS systems may adopt, including a long synchronisation period at the start of the transmission to ensure correct clock recovery.

In figure 8.13 is a similar approach is shown that was proposed in [71] for both V2I and V2V scenarios. Theoretical relationships were defined for the angles of irradiance derived, related to those in chapter 3. The cooperative scheme presented in [71] adds a dimension to the previous report since the source transmits to all interested parties who have LOS and doesn't necessarily rely on the relay of transmission unless necessary.

The main issue with the previous report is that the SNR is constantly decreasing because of the cascading receiver noise and increasingly uncertain demodulation confidence. Therefore, if there is a clear LOS between a transmitter traffic light and several receivers then there may be sufficient SNR to support direct transmission without the necessity of a relay. The added dimension in [71] is the introduction of MRC that allows secondary and tertiary (i.e. not the closest) receivers to decide which signal they want to receive between the source itself or the vehicle in front, which also automatically relays the same information.

Using this method, it is immediately clear from the measurement of BER results that cooperation adds a significant degree of reliability to the overall performance of the system. In terms of the V2V approach, figure 8.14 clearly shows that the BER is decreased by approximately three orders of magnitude by introducing the MRC approach across all distances up to 55 m, which can either be translated to a reliability gain, which is of the utmost importance in ITS systems, or into an SNR gain, where a higher data rate could be used to transmit more information. Likewise, for a V2I configuration there is a maximum of four orders of magnitude gain between the two systems as illustrated in figure 8.15.

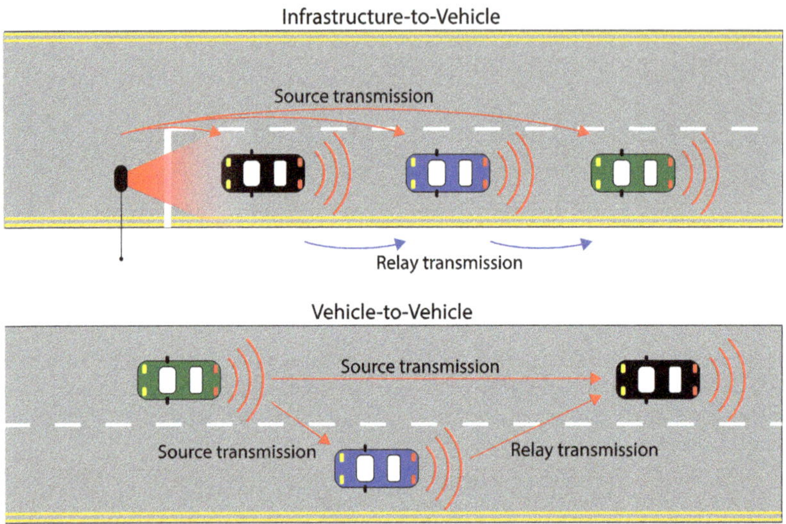

Figure 8.13. The systems tested in [71] were platoons consisting of both infrastructure-to-vehicle and vehicle-to-vehicle. The advantages of these approaches are that there is a contingency if one source-to-receiver link was to fail, the receiver would still receive the information from a secondary or tertiary source, improving reliability of the overall system in a cooperative manner.

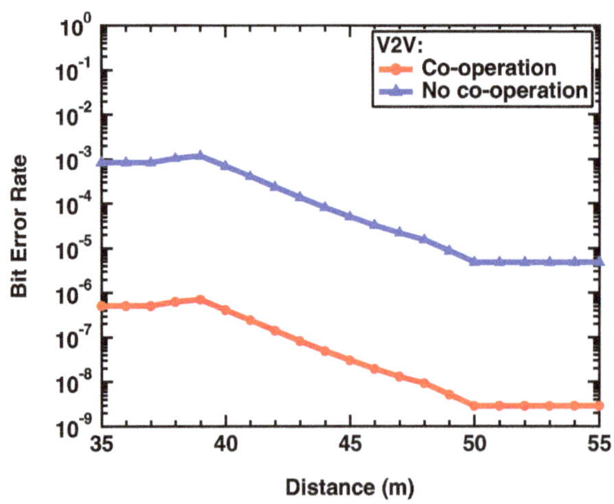

Figure 8.14. The BER of the cooperative system reported in [71] for vehicle-to-vehicle links shows around three orders of magnitude improvement over the normal transmit and re-transmit approach shown previously in figure 8.11.

8.3.2 Platooning

The concept of re-transmitting information to the vehicles behind the leader is called platooning and groups of vehicles can be referred to as platoons [72–74]. Interference in VLC-based ITS is an important consideration due to the large optical footprint that can straddle several lanes, as illustrated in figure 8.16 as the

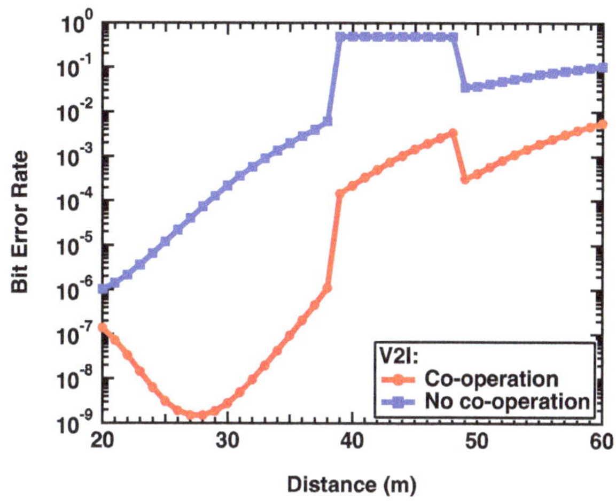

Figure 8.15. Similar BER improvements can be observed for vehicle-to-infrastructure links, with several orders of magnitude improvement measured over all distances between vehicles using the cooperative approach reported in [71].

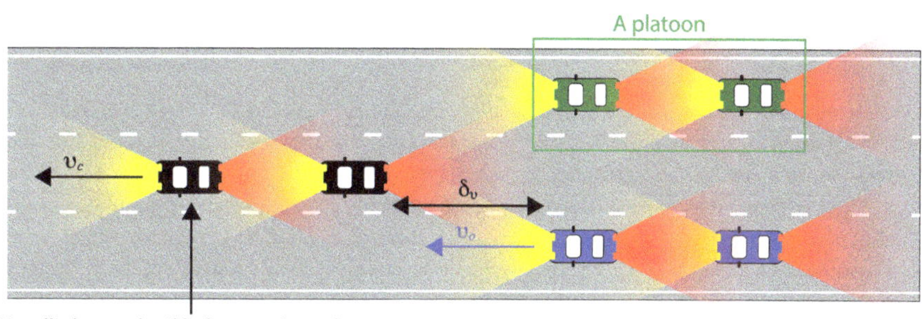

Figure 8.16. One of the issues with vehicle headlines is that they are not well-directed links by their nature; they are meant to be seen for obvious reasons. Therefore, the optical footprints can straddle more than one lane of a road which can be an issue as illustrated here.

information shared in one platoon may be unnecessary for others. For safe operation, platoons can tolerate a maximum latency of 2–300 ms [75] and as a result VLC is a good choice. If the scenario in figure 8.16 is considered, a scenario may occur where the middle platoon is going to overtake those on either side of it. The distance between the platoons is given by δ_d and the relative velocities of the centre and outside platoons are υ_c and υ_o, respectively. The relative velocity at which the centre platoon overtakes is given as $\delta_\upsilon = \upsilon_c - \upsilon_o$.

Interference may occur in two scenarios; firstly when cars pass each other in different lanes and secondly when platoons are taking a corner, as is highlighted in figure 8.17, where direct LOS paths are introduced between the lead vehicle and its followers which previously did not exist, thus introducing a multi-path effect and degrading the signal quality. In the work reported in [73], a simulation was performed that took into account a platoon of 8 vehicles that are initially a distance

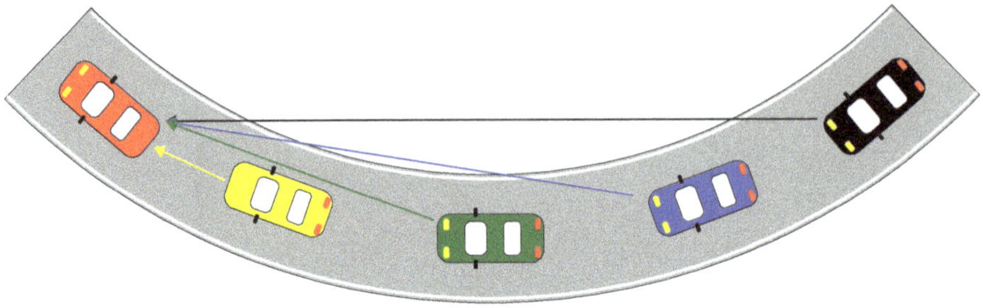

Figure 8.17. Another source of interference for platoons, aside from inter-lane interference is when cornering, the headlight angle of emission may interact with the receivers more than one car ahead in the platoon, which can interrupt proper transmission between the platoon and impact safety.

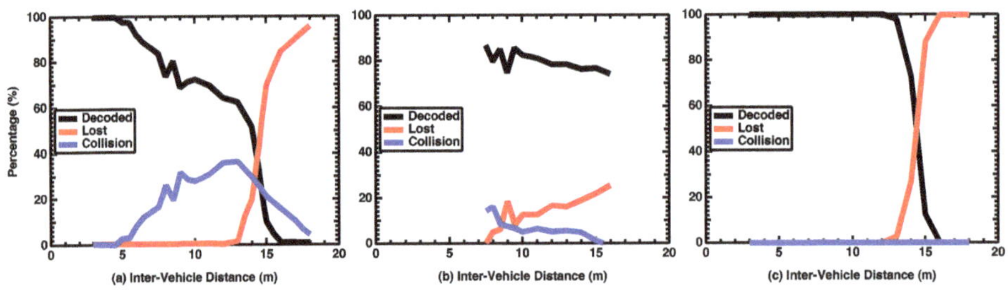

Figure 8.18. The relative packet success when considering interference between lanes; (a) when there is a small inter-vehicle distance, i.e. <5 m, no packets are dropped; however, when this distance increases, the number of packets lost or collided (resulting in retransmission requests) increases significantly until all the packets are lost; (b) when travelling around a curved road, the results are much better over long distances, and the packets can be recovered >75% of the time, while finally (c) when highly directional matrices of light are used, packets can always be recovered for reasonable distances due to the lower interference and higher optical power received.

of 10 m behind the outside platoons, with a relative velocity difference of 10 m/s. Each LED has an angle of emission of 8° and transmits at a rate of 1 Mb/s with an assumed attenuation of 60 dB.

It is shown in [73] and in figure 8.18(a) that when the inter-vehicle distance in each platoon is 5 m there are no packets dropped in the platoon. When the distance increases beyond 5 m, the packet loss increases and the system eventually fails when single LED chips are used for the transmission. On the other hand, when travelling around a curved road (radius 230 m), the system never fails due to the ever-present LOS availability as mentioned previously; however, due to a reduction in the SNR caused by the multi-path effect, transmission can never occur without packets being dropped (figure 8.18(b)). Finally, when matrices of LEDs are used, the additional footprint introduced enables a more reliable communication and distances of up to 12 m can be supported without any packets dropped on a straight road (figure 8.18(c)).

8.3.3 Weather

One of the key considerations in outdoor communication systems such as FSO is commonly the atmospheric conditions present when transmission occurs. Due to the relative size of atmospheric particles in comparison to that of the optical wavelengths, effects such as scintillation, scattering and turbulence become important [76–78]. The report in [78] studies this in detail by evaluating the communication distances that can be achieved in different weather conditions.

The system presented considers the scenario shown in figure 8.19, where there is a direct link between two vehicles that is perturbed by one interferer. The distance between transmitter and receiver is assumed to vary between 3 and 10 m. The bandwidth of the system was set at 2 MHz and the luminous efficacy of the LEDs in the headlights was 250.3 lm/W. A number of weather conditions were considered, including light and dense fog, and their visibility parameters were 0.1 and 0.05 km, respectively. The attenuation coefficients of each weather condition considered are stated in table 8.3, where V indicates visibility [79].

The impact of these weather conditions can be inferred from figure 8.20 where it is obviously clear that rain and wet snow have similar attenuation on the optical beam, but other weather conditions severely limit the available distances of the link. The data rate was set to 3 Mb/s at a 15 dB SINR. The maximum achievable distances between the two vehicles under the different weather conditions are outlined in table 8.4.

As is illustrated in table 8.4, the nominal distance can be approached regardless of weather conditions, with the most detrimental conditions being dense fog and dry snow, as has been widely reported in the literature for FSO systems [39, 41, 78, 79].

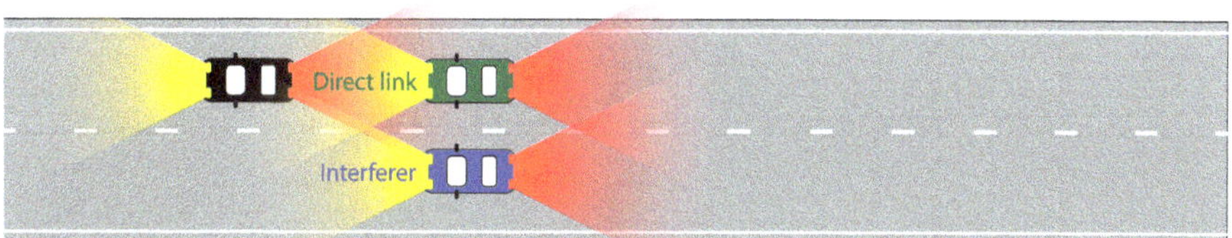

Figure 8.19. The impact of weather on the links when considering interferers is also an important aspect when it comes to safety. The scenario tested in [78] is illustrated here, where there is an interferer considered for inter-vehicle different distances and weather conditions.

Table 8.3. Typical attenuation values for different weather conditions.

Weather condition	Attenuation (dB/km)
Dense fog (V = 0.05 km)	78.8
Light fog (V = 0.1 km)	39.4
Rain (rate = 90 mm h^{-1})	21.9
Wet snow (rate = 10 mm h^{-1})	19.9
Dry snow (rate = 10 mm h^{-1})	131

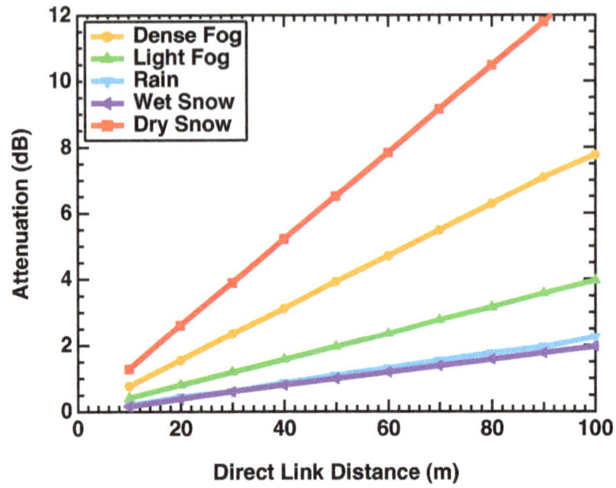

Figure 8.20. Attenuation values for different weather conditions over varying link distances.

Table 8.4. Achievable distances that the link can operate over considering a variable distance interferer and different weather conditions reported in [79].

Nominal achievable distance (m)	Interferer distance (m)	Dense fog	Light fog	Rain	Wet snow	Dry snow
60	3	45	55	55	55	40
80	4.5	60	70	75	70	50
90	7	70	80	85	80	55
100	8.5	80	100	100	100	70
100	10	90	100	100	100	75

8.4 Summary

In this chapter, IoT and ITS were discussed. In the opinion of the author, these two subjects go hand-in-hand in the 'everything connected' paradigm. One of the most interesting aspects of IoT is the inherent ability of VLC to harvest the dc left over from the LED biasing. If this energy is not recovered, it is simply wasted and could otherwise be used to power receiver circuits or store energy for other applications. Several reports were published recently that are covered in this work in terms of recovering the ac and dc components. It was shown for both inorganic and organic devices that energy harvesting can be performed whilst still recovering information successfully. Other methods for energy recovery were shown, with lens, mirror and reflector configurations shown.

In terms of ITSs, a number of systems were discussed from the literature that look at different aspects. Platooning was considered including its safety aspects and traffic light-to-car communications were considered. It was shown clearly that BER values could be improved through cooperative systems. Inter-lane interference was also considered as well as weather conditions.

References

[1] Atzori L, Iera A and Morabito G 2010 The internet of things: a survey *Comput. Netw.* **54** 2787–805

[2] Wang J, Al-Kinani A, Zhang W and Wang C-X 2017 A new VLC channel model for underground mining environments *2017 13th Int. Wireless Communications and Mobile Computing Conf. (IWCMC)* (Piscataway, NJ: IEEE) pp 2134–9

[3] Chen M, Wan J, Gonzalez S, Liao X and Leung V C M 2013 A survey of recent developments in home M2M networks *IEEE Commun. Surv. Tutor* **16** 98–114

[4] Nizzi F *et al* 2019 Data dissemination to vehicles using 5G and VLC for smart cities *2019 AEIT Int. Annual Conf. (AEIT)* (Piscataway, NJ: IEEE) pp 1–5

[5] Tiwari S V, Sewaiwar A and Chung Y-H 2015 Smart home technologies using visible light communication *2015 IEEE Int. Conf. on Consumer Electronics (ICCE)* (Piscataway, NJ: IEEE) pp 379–80

[6] Kumar A, Mihovska A, Kyriazakos S and Prasad R 2014 Visible light communications (VLC) for ambient assisted living *Wirel. Pers. Commun.* **78** 1699–717

[7] Singh A, Rehman S U, Yongchareon S and Chong P H J 2020 Sensor technologies for fall detection systems: a review *IEEE Sens. J.* **20** 6889–919

[8] Armstrong J, Sekercioglu Y A and Neild A 2013 Visible light positioning: a roadmap for international standardization *IEEE Commun. Mag.* **51** 68–73

[9] Lochner C M, Khan Y, Pierre A and Arias A C 2014 All-organic optoelectronic sensor for pulse oximetry *Nat. Commun.* **5** 1–7

[10] Han D, Khan Y, Ting J, Zhu J, Combe C, Wadsworth A, McCulloch I and Arias A C 2020 Pulse oximetry using organic optoelectronics under ambient light *Adv. Mater. Technol.* **5** 1901122

[11] Abadi M M, Hazdra P, Bohata J, Chvojka P, Haigh P A, Ghassemlooy Z and Zvanovec S 2020 A head/taillight featuring hybrid planar visible light communications/millimetre wave antenna for vehicular communications *IEEE Access* 1

[12] Yang H, Alphones A, Zhong W-D, Chen C and Xie X 2019 Learning-based energy-efficient resource management by heterogeneous RF/VLC for ultra-reliable low-latency industrial IoT networks *IEEE Trans. Industr. Inform.* **16** 5565–76

[13] Xu L D, He W and Li S 2014 Internet of things in industries: a survey IEEE Trans. Industr *IEEE Trans. Industr. Inform.* **10** 2233–43

[14] Al-Fuqaha A, Guizani M, Mohammadi M, Aledhari M and Ayyash M 2015 Internet of things: a survey on enabling technologies, protocols, and applications *IEEE Commun. Surv. Tutor.* **17** 2347–76

[15] Ramos J S, Demirkol I, Paradells J, Vössing D, Gad K M and Kasemann M 2016 Towards energy-autonomous wake-up receiver using visible light communication *2016 13th IEEE Annual Consumer Communications & Networking Conf. (CCNC)* (Piscataway, NJ: IEEE) pp 544–9

[16] Kahn J M and Barry J R 1997 Wireless infrared communications *Proc. IEEE* **85** 265–98

[17] Haigh P A, Ghassemlooy Z, Le Minh H, Rajbhandari S, Arca F, Tedde S F, Hayden O and Papakonstantinou I 2012 Exploiting equalization techniques for improving data rates in organic optoelectronic devices for visible light communications *J. Lightwave Technol.* **30** 3081–8

[18] Wang Z, Tsonev D, Videv S and Haas H 2015 On the design of a solar-panel receiver for optical wireless communications with simultaneous energy harvesting *IEEE J. Sel. Areas Commun.* **33** 1612–23

[19] Zhang S, Tsonev D, Videv S, Ghosh S, Turnbull G A, Samuel I D W and Haas H 2015 Organic solar cells as high-speed data detectors for visible light communication *Optica* **2** 607–10

[20] Phang J C H, Chan D S H and Phillips J R 1984 Accurate analytical method for the extraction of solar cell model parameters *Electron. Lett.* **20** 406–8

[21] Reddy P J 2010 *Science Technology of Photovoltaics* (Hyderabad: Book Syndicate Publications)

[22] Kassem A and Darwazeh I 2019 Exploiting negative impedance converters to extend the bandwidth of LEDs for visible light communication *2019 26th IEEE Int. Conf. on Electronics, Circuits and Systems (ICECS)* (Piscataway, NJ: IEEE) pp 302–5

[23] Fakidis J, Ijaz M, Kucera S, Claussen H and Haas H 2014 On the design of an optical wireless link for small cell backhaul communication and energy harvesting *2014 IEEE 25th Annual Int. Symp. on Personal, Indoor, and Mobile Radio Communication (PIMRC)* (Piscataway, NJ: IEEE) pp 58–62

[24] Shin W-H, Yang S-H, Kwon D-H and Han S-K 2016 Self-reverse-biased solar panel optical receiver for simultaneous visible light communication and energy harvesting *Opt. Express* **24** A1300–5

[25] Obeed M, Dahrouj H, Salhab A M, Zummo S A and Alouini M-S 2019 DC-bias and power allocation in cooperative VLC networks for joint information and energy transfer *IEEE Trans. Wireless Commun.* **18** 5486–99

[26] Obeed M, Dahrouj H, Salhab A M, Zummo S A and Alouini M-S 2018 DC-bias allocation in cooperative VLC networks via joint information and energy transfer *2018 IEEE Global Communications Conf. (GLOBECOM)* (Piscataway, NJ: IEEE) pp 1–6

[27] Kaur K and Rampersad G 2018 Trust in driverless cars: investigating key factors influencing the adoption of driverless cars *J. Eng. Technol. Manage.* **48** 87–96

[28] Bradshaw-Martin H and Easton C 2014 Autonomous or 'driverless' cars and disability: a legal and ethical analysis *Eur. J. Curr. Leg. Issues* **20**

[29] Jones L 2017 Driverless cars: when and where? *Eng. Technol.* **12** 36–40

[30] Hussain R and Zeadally S 2018 Autonomous cars: research results, issues, and future challenges *IEEE Commun. Surv. Tutor.* **21** 1275–313

[31] Cerruela García G, Luque Ruiz I and Gómez-Nieto M Á 2016 State of the art, trends and future of bluetooth low energy, near field communication and visible light communication in the development of smart cities *Sensors* **16** 1968

[32] Cailean A-M and Dimian M 2017 Impact of IEEE 802.15. 7 standard on visible light communications usage in automotive applications *IEEE Commun. Mag.*

[33] Masini B M, Bazzi A and Zanella A 2018 A survey on the roadmap to mandate on board connectivity and enable V2V-based vehicular sensor networks *Sensors* **18** 2207

[34] World Health Organization 2015 *Global Status Report on Road Safety 2015* (World Health Organization)

[35] Ameratunga S, Hijar M and Norton R 2006 Road-traffic injuries: confronting disparities to address a global-health problem *Lancet* **367** 1533–40

[36] World Health Organization *et al* 2017 The top 10 causes of death: World Health Organization (updated January 2017; cited February 9, 2017) *World Health Organization Fact Sheets* (http://www.who.int/mediacentre/factsheets/fs310/en)

[37] Najm W G *et al* 2010 Frequency of target crashes for intellidrive safety systems *Technical Report* National Highway Traffic Safety Administration
[38] Sandalidis H G, Tsiftsis T A, Karagiannidis G K and Uysal M 2008 BER performance of FSO links over strong atmospheric turbulence channels with pointing errors *IEEE Commun. Lett.* **12** 44–6
[39] Tsiftsis T A, Sandalidis H G, Karagiannidis G K and Uysal M 2009 Optical wireless links with spatial diversity over strong atmospheric turbulence channels *IEEE Trans. Wireless Commun.* **8** 951–7
[40] Xu G, Zhang X, Wei J and Fu X 2004 Influence of atmospheric turbulence on FSO link performance *Proc. SPIE* **5281** 816–23
[41] Popoola W O and Ghassemlooy Z 2009 BPSK subcarrier intensity modulated free-space optical communications in atmospheric turbulence *J. Lightwave Technol.* **27** 967–73
[42] Zvánovec S, Žák P, Chvojka P, Kudláček I, Haigh P A and Ghassemlooy Z 2017 Visible light communications based on street lighting *Visible Light Communications: Theory and Applications* 283
[43] Cailean A-M and Dimian M 2017 Current challenges for visible light communications usage in vehicle applications: a survey *IEEE Commun. Surv. Tutor* **19** 2681–703
[44] Uysal M, Miramirkhani F, Narmanlioglu O, Baykas T and Panayirci E 2017 IEEE 802.15. 7r1 reference channel models for visible light communications *IEEE Commun. Mag.* **55** 212–7
[45] Jiang D and Delgrossi L 2008 IEEE 802.11 p: Towards an international standard for wireless access in vehicular environments *VTC Spring 2008-IEEE Vehicular Technology Conf.* (Piscataway, NJ: IEEE) pp 2036–40
[46] Atallah R, Khabbaz M and Assi C 2016 Multihop V2I communications: a feasibility study, modeling, and performance analysis IEEE Trans *Vehicular Technol* **66** 2801–10
[47] Yao Y, Rao L and Liu X 2013 Performance and reliability analysis of IEEE 802.11 p safety communication in a highway environment *IEEE Trans. Vehicular Technol.* **62** 4198–212
[48] Bilstrup K, Uhlemann E, Ström E and Bilstrup U 2009 On the ability of the 802.11 p MAC method and STDMA to support real-time vehicle-to-vehicle communication *EURASIP J. Wireless Commun. Netw.* **2009** 902414
[49] Eichler S 2007 Performance evaluation of the IEEE 802.11 p WAVE communication standard *2007 IEEE 66th Vehicular Technology Conf.* (Piscataway, NJ: IEEE) pp 2199–203
[50] Fries B, Gut C, Laudenbach T and Mühlmeier M 2014 Laser light for the Audi R18 e-tron quattro racing car *ATZ Worldwide* **116** 24–7
[51] Ulrich L 2013 Whiter brights with lasers *IEEE Spectr* **50** 36–56
[52] Christnacher F, Poyet J-M, Laurenzis M, Moeglin J-P and Taillade F 2010 Bistatic range-gated active imaging in vehicles with LEDs or headlights illumination *Proc. SPIE* **7675** 76750J
[53] Ji P, Tsai H-M, Wang C and Liu F 2014 Vehicular visible light communications with LED taillight and rolling shutter camera *2014 IEEE 79th Vehicular Technology Conf. (VTC Spring)* (Piscataway, NJ: IEEE) pp 1–6
[54] Yamazato T *et al* 2014 Image-sensor-based visible light communication for automotive applications *IEEE Commun. Mag.* **52** 88–97
[55] Takai I, Harada T, Andoh M, Yasutomi K, Kagawa K and Kawahito S 2014 Optical vehicle-to-vehicle communication system using LED transmitter and camera receiver *IEEE Photonics J.* **6** 1–14

[56] Takai I, Ito S, Yasutomi K, Kagawa K, Andoh M and Kawahito S 2013 LED and CMOS image sensor based optical wireless communication system for automotive applications *IEEE Photonics J.* **5** 6801418

[57] Goto Y, Takai I, Yamazato T, Okada H, Fujii T, Kawahito S, Arai S, Yendo T and Kamakura K 2016 A new automotive VLC system using optical communication image sensor *IEEE Photonics J.* **8** 1–17

[58] Wang C, Liu Y, Li W, Wang F and Chi N 2019 A 375Mb/s real-time internet of vehicles system based on automotive headlight utilizing OFDM-64QAM modulation format *2019 18th Int. Conf. on Optical Communications and Networks (ICOCN)* (Piscataway, NJ: IEEE) pp 1–3

[59] Cailean A-M, Cagneau B, Chassagne L, Dimian M and Popa V 2015 Novel receiver sensor for visible light communications in automotive applications *IEEE Sens. J.* **15** 4632–9

[60] Saito T, Haruyama S and Nakagawa M 2008 A new tracking method using image sensor and photo diode for visible light road-to-vehicle communication *2008 10th Int. Conf. on Advanced Communication Technology* vol 1 (Piscataway, NJ: IEEE) pp 673–8

[61] Okada S, Yendo T, Yamazato T, Fujii T, Tanimoto M and Kimura Y 2009 On-vehicle receiver for distant visible light road-to-vehicle communication *2009 IEEE Intelligent Vehicles Symp.* (Piscataway, NJ: IEEE) pp 1033–8

[62] Cailean A, Cagneau B, Chassagne L, Topsu S, Alayli Y and Blosseville J-M 2012 Visible light communications: application to cooperation between vehicles and road infrastructures *2012 IEEE Intelligent Vehicles Symp.* (Piscataway, NJ: IEEE) pp 1055–9

[63] Lourenço N, Terra D, Kumar N, Alves L N and Aguiar R L 2012 Visible light communication system for outdoor applications *2012 8th Int. Symp. on Communication Systems, Networks & Digital Signal Processing (CSNDSP)* (Piscataway, NJ: IEEE) pp 1–6

[64] Kumar N, Lourenço N, Terra D, Alves L N and Aguiar R L 2012 Visible light communications in intelligent transportation systems *2012 IEEE Intelligent Vehicles Symp.* (Piscataway, NJ: IEEE) pp 748–53

[65] Terra D, Kumar N, Lourenço N, Alves L N and Aguiar R L 2011 Design, development and performance analysis of DSSS-based transceiver for VLC *2011 IEEE EUROCON-Int. Conf. on Computer as a Tool* (Piscataway, NJ: IEEE) pp 1–4

[66] Cailean A-M, Cagneau B, Chassagne L, Topsu S, Alayli Y and Dimian M 2013 Visible light communications cooperative architecture for the intelligent transportation system *2013 IEEE 20th Symp. on Communications and Vehicular Technology in the Benelux (SCVT)* (Piscataway, NJ: IEEE) pp 1–5

[67] Corsini R, Pelliccia R, Cossu G, Khalid A M, Ghibaudi M, Petracca M, Pagano P and Ciaramella E 2012 Free space optical communication in the visible bandwidth for V2V safety critical protocols *2012 8th Int. Wireless Communications and Mobile Computing Conf. (IWCMC)* (Piscataway, NJ: IEEE) pp 1097–102

[68] Li G, Hu F, Zhao Y and Chi N 2019 Enhanced performance of a phosphorescent white LED CAP 64QAM VLC system utilizing deep neural network (DNN) post equalization *2019 IEEE/CIC Int. Conf. on Communications in China (ICCC)* (Piscataway, NJ: IEEE) pp 173–6

[69] Agarwal A and Little T D C 2010 Role of directional wireless communication in vehicular networks *2010 IEEE Intelligent Vehicles Symp.* (Piscataway, NJ: IEEE) pp 688–93

[70] Cailean A, Cagneau B, Chassagne L, Topsu S, Alayli Y and Dimian M 2013 A robust system for visible light communication *2013 IEEE 5th Int. Symp. on Wireless Vehicular Communications (WiVeC)* (Piscataway, NJ: IEEE) pp 1–5

[71] Cui Z, Yue P and Ji Y 2016 Study of cooperative diversity scheme based on visible light communication in VANETs *2016 Int. Conf. on Computer, Information and Telecommunication Systems (CITS)* (Piscataway, NJ: IEEE) pp 1–5

[72] Rajamani R, Tan H-S, Law B K and Zhang W-B 2000 Demonstration of integrated longitudinal and lateral control for the operation of automated vehicles in platoons *IEEE Trans. Control Syst. Technol.* **8** 695–708

[73] Schettler M, Memedi A and Dressler F 2019 The chosen one: combating VLC interference in platooning using matrix headlights *2019 IEEE Vehicular Networking Conf. (VNC)* (Piscataway, NJ: IEEE) pp 1–4

[74] Robinson T, Chan E and Coelingh E 2010 Operating platoons on public motorways: An introduction to the SARTRE platooning programme *17th World Congress on Intelligent Transport Systems* vol 1 p 12

[75] Segata M, Bloessl B, Joerer S, Sommer C, Gerla M, Cigno R L and Dressler F 2015 Toward communication strategies for platooning: simulative and experimental evaluation *IEEE Trans. Vehicular Technol.* **64** 5411–23

[76] Gern A, Moebus R and Franke U 2002 Vision-based lane recognition under adverse weather conditions using optical flow *Intelligent Vehicle Symp., 2002. IEEE* vol 2 (Piscataway, NJ: IEEE) pp 652–7

[77] Nadeem F, Kvicera V, Awan M S, Leitgeb E, Muhammad S S and Kandus G 2009 Weather effects on hybrid FSO/RF communication link *IEEE J. Sel. Areas Commun.* **27** 1687–97

[78] Singh G, Srivastava A and Bohara V A 2019 Impact of weather conditions and interference on the performance of VLC based V2V communication *2019 21st Int. Conf. on Transparent Optical Networks (ICTON)* (Piscataway, NJ: IEEE) pp 1–4

[79] Singh G, Srivastava A and Bohara V A 2018 On feasibility of VLC based car-to-car communication under solar irradiance and fog conditions *Proc. 1st Int. Workshop on Communication and Computing in Connected Vehicles and Platooning* pp 1–7

Chapter 9

Oxygen sensing

Nikolaos Salaris[1,2] and Paul Anthony Haigh[3]
[1]*Department of Electronic and Electrical Engineering, University College London, London, United Kingdom*
[2]*Department of Mechanical Engineering, University College London, London, United Kingdom*
[3]*School of Agriculture, Engineering and Mathematics, Newcastle University, Newcastle-upon-Tyne, United Kingdom*

9.1 Introduction

In recent decades, as technology has advanced and electronics have become miniaturised, the door for on/in-body monitoring and connected health has opened. This leads to improved patient care linkage and quality of life. One of the major success stories of on-body monitoring is glucose monitoring as an early warning detection for diabetes. Over the years, photonic biosensors have developed rapidly, driven by the miniaturisation of electronic circuits.

One of the newest and most interesting applications that has emerged for VLC is healthcare sensing. Several technological domains have begun adopting VLC in applications ranging from pulse oximetry in soft tissue using PLEDs [1] through to infectious disease detection and quantification by imaging of microfluidic tests [2] and even x-ray detection using OPDs [3], among others. When combined with real-time monitoring and data analytics, diagnosis and treatment of illnesses can be accelerated and constantly monitored, leading to personalised care, as opposed to the one-size-fits-all approach traditionally deployed.

In the opinion of the authors, this is one of the most exciting emerging areas for VLC, since in the literature optical systems are not optimised in many cases. There are significant improvements that can be made, leading to active contributions within the healthcare research community, and improved quality of care linkage to patients, particularly in developing nations where specialised equipment is often scarcely available. One of the key reasons that photonic biosensors are becoming an area of significant interest is because they are free from RF components, which are restricted in certain circumstances, such as magnetic resonance imaging. As well as

doi:10.1088/978-0-7503-1680-4ch9

the enabling technology for photonic biosensors, VLC can also be used to transfer information transdermally, meaning existing permanent fixtures such as pacemakers can be maintained in place; however, rather than transmitting the information by a RF carrier, the information is instead transmitted on visible light.

In this chapter, the oxygen sensing application for digital healthcare linkage with VLC will be discussed. In this approach, visible (and NIR) light is used to measure the quantities of oxygen in a given medium, such as blood [4, 5] and tissue [6, 7].

9.2 Applications

9.2.1 Cancer studies

It is fundamental that oxygen sensors are readily available, accurate and low cost, due to the sheer volume of potential measurements and tests that can be performed. Therefore, there has been an abundance of literature produced in the fields across numerous domains, from biochemistry [8] to electronic engineering [9, 10].

One of the most important applications evolves around cancer studies and more specifically in the investigation of tumour growth. Hypoxia strongly relates to the growth or recession of a solid tumour in response to traditional cancer therapies, while oxygenation of the surrounding areas can be considered a prognostic factor [11, 12]. The amount of oxygen present in the tumourous and surrounding tissue varies significantly from case to case, which clearly indicates the importance of developing personalised oxygen sensors [13]. This is further underlined by the fact that there is no clear correlation between oxygen quantities present at the primary tumour growth site and the growth of secondary tumours elsewhere in the body (metastasis), although oxygen distribution appears to be consistent [14].

In general, hypoxia occurs when oxygen demand outstrips supply within functional tissue and there are two main causes: (i) limited distribution because of the spatial geometry of the cells [15] and (ii) limited oxygen delivery due to restricted flow of red blood cells [16]. Hypoxia can trigger genetic changes that promote tumour growth due to cells attempting to adapt to the reduced oxygen level [17]. Therefore, monitoring both blood flow and local tissue oxygenation is extremely important to properly study cancer cell growth *in vivo*. As such, new approaches that utilise phosphorescent quenching have been applied for imaging of tumours in a non-invasive manner [16, 18].

9.2.2 Cell cultures

Artificial tissue growth is required to improve patient outcome in tissue replacement therapies as used in numerous medical areas, and has resulted in significant advances in areas such as 3D bio-printing for a small number of tissue types [19, 20]. One of the most important aspects of oximetry with respect to tissue replacement therapies is to closely relate oxygen measurements to the core development of the tissue [21], in characterisations such as the cell number, biochemical composition, vascularisation and diffusion properties, etc, which define the success of the transplant [22, 23].

In respirometry, oxygen consumption has been the focus of enormous interest during the last few decades due to the role that oxygen plays in cell metabolism, where tissue relies on the supply of sufficient oxygen levels for its continuous growth.

In general, oxygen content of cell cultures can be measured in numerous ways such as with electrical/chemical methods [24, 25], all of which have faced challenges, including spatial resolution and cell damage. In general, the least invasive approach has been through lightwave systems that exploit phosphorescence quenching [26], and studies have demonstrated oxygen distribution and correlation with the aforementioned parameters [21]. These have led to mathematical models that have been able to describe nutrient transport and cell growth within bioreactors [27]. Finally, hypoxic conditions are generally time-dependent due to fluctuations in oxygen availability and lightwave systems allow real-time monitoring [21].

9.3 Types of oximetry

9.3.1 Haemoglobin-based oximetry

Here, the focus is on two-wavelength absorptiometric pulse oximetry. The optical system is generally established in either a transmissive, i.e. passing through the tissue and blood, or reflective mode. The different layers of tissue are highly scattering media with uniform back scatter [4–7], see figure 9.1. At the receiver, the quantity of optical power recovered is proportional to the absorbance of the medium [1], which can be related to the ratio of de-oxygenated and oxygenated haemoglobin via the Beer–Lambert law [28] and thus establish the oxygen saturation levels. The dc part of the signal is dependent on the tissue, venous and non-pulsating blood oxygenation while the ac part of the signal is linked to the pulsatile variations. The Beer–Lambert law is given by [28]:

$$T_{BL} = e^{-A} \tag{9.1}$$

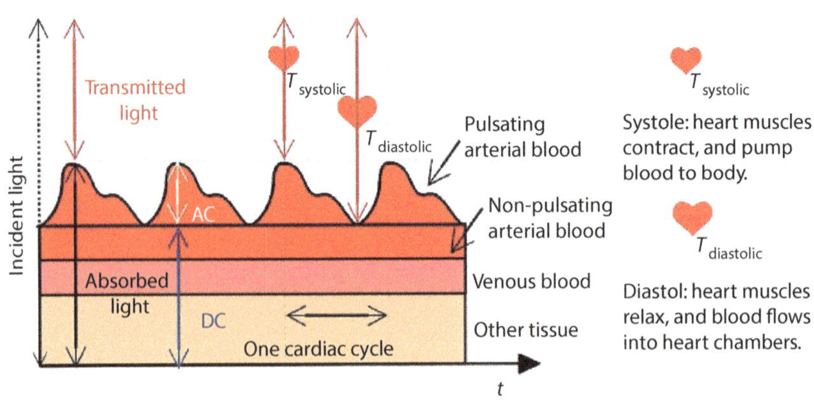

Figure 9.1. Schematic depicting the absorption of light from the various layers of skin and the method of pulse and arterial blood extraction [1]. Reprinted with permission from Springer Nature: Nature Communications [1] © (2014).

where T_{BL} is the transmittance and A is the absorbance. The transmittance is given by [29]:

$$T_{BL} = \frac{I}{I_0} \tag{9.2}$$

where I and I_0 are the transmitted intensity and initial incident intensity, respectively. The absorbance is given by [29]:

$$A = lC\varepsilon \tag{9.3}$$

where l is the path length and ε is the molar absorption coefficient. Absorption peaks when the heart contracts and troughs when it relaxes, corresponding to a common, and more familiar, electrocardiogram that monitors the heart rate for irregularities. Within these two phases of the heart cycle, two types of haemoglobin (oxyhaemoglobin and deoxyhaemoglobin) are present and they have different absorptivities across different wavelengths [30]. Thus, different wavelengths (commonly red and IR) are used in the sensor (refer to figure 9.2).

Blood oxygenation can be quantified in proportion to the oxygen saturation SO_2 as follows [30]:

$$SO_2 = \frac{CHbO_2}{CHbO_2 + CHb} \tag{9.4}$$

where O_2 is oxygen, Hb is deoxyhaemoglobin, HbO_2 is oxyhaemoglobin and C is the haemoglobin concentration.

Using the Beer–Lambert law (9.1), the ratio R_{os} between the absorbance of red ($A_{\lambda 1}$) and IR ($A_{\lambda 2}$) can be found:

$$R_{os} = \frac{A_{\lambda 1}}{A_{\lambda 2}} = \frac{\ln(T_1)}{\ln(T_2)} \tag{9.5}$$

Thus, via the molar absorptivity of two wavelengths, the oxygen saturation becomes [31]:

$$SO_2 = \frac{\varepsilon \mathrm{Hb}(\lambda_1) - \varepsilon \mathrm{Hb}(\lambda_2 R_{os})}{\varepsilon \mathrm{Hb}(\lambda_1) - \varepsilon \mathrm{HbO_2}(\lambda_1) + R_{os}[\varepsilon \mathrm{HbO_2}(\lambda_2) - \varepsilon \mathrm{Hb}(\lambda_2)]} \tag{9.6}$$

The optical path lengths for both wavelengths are assumed to be equal, however, the refractive indices of the path transversed have a wavelength dependence, and hence are different for each wavelength [31]:

$$l \longrightarrow \mathrm{DPF}(\lambda) \tag{9.7}$$

where DPF is the differential path length factor, and as such, a modified R_{os} is calculated to provide higher accuracy (refer to [31]).

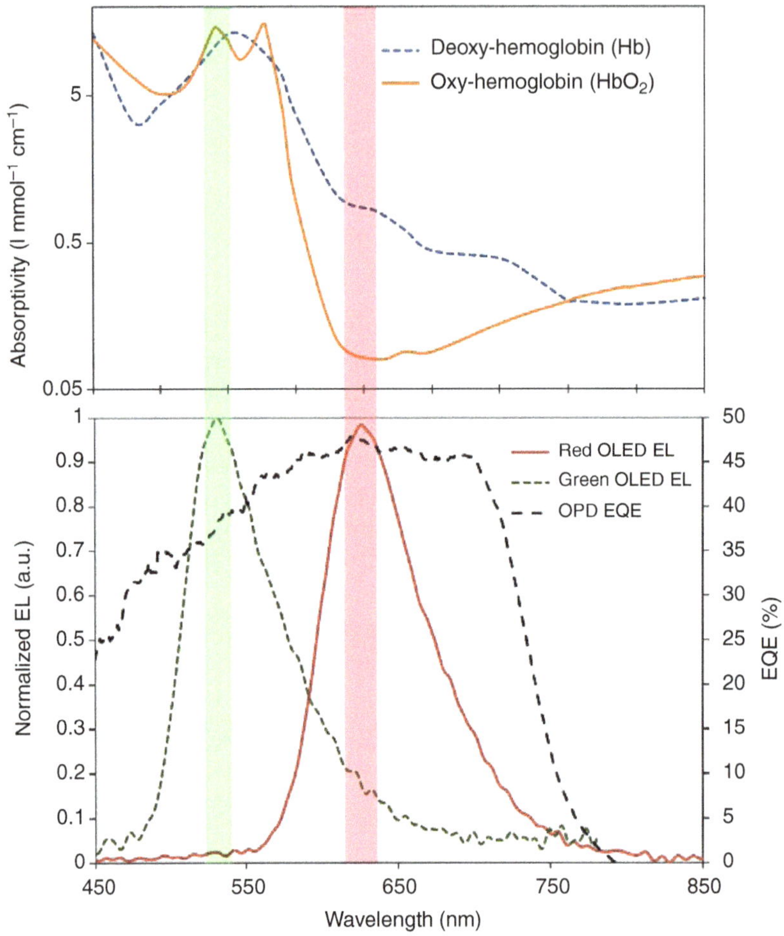

Figure 9.2. (Top) Absorptivity of oxygenated (orange solid line) and de-oxygenated (blue dashed line) haemoglobin in arterial blood as a function of wavelength. The wavelengths corresponding to the peak OLED electroluminescence spectra are highlighted to show that there is a difference in deoxy- and oxy-haemoglobin molar absorption coefficient at the wavelengths of interest. (Bottom) OPD external quantum efficiency (black dashed line) at short circuit, and normalised electroluminescence spectra of red (red solid line) and green (green dashed line) OLEDs [1]. Reprinted with permission from Springer Nature: Nature Communications [1] © (2014).

9.3.2 Phosphorescence quenching

Miniaturised optical sensors have been successfully developed based on the relative intensity measurements described above [1, 32]. However, the literature has clearly demonstrated the numerous limitations of such a methodology in terms of clinical outcome [33, 34]. To add to that, the use of phosphorescence quenching results in improved sensitivities [35], providing absolute measurements even when haemoglobin is not present [36]. Presuming rudimentary knowledge of photoluminescence and fluorescence (refer to [35]), phosphorescent quenching occurs when an intermolecular interaction between the luminophore and a quencher results in non-radiative deactivation of the excited state. This leads to a phenomenon where the phosphorescent light emitted varies in intensity and lifetime depending on the presence (and quantity) of oxygen [37, 38].

It must also be stated that although the quantification of such variations are done with remarkable accuracy, the underlying mechanism controlling the interaction between the two molecules is uncertain. The most strongly assumed process is dynamic collisions. In which the result of the collision is the inter-system crossing of the luminophore to the ground energy state (i.e. radiative decay and photon emission after a Stoke's shift) and the simultaneous production of a higher state oxygen and can be described by [39]:

$$L^* + O_2 = O_2^* + L + \text{radiation} \tag{9.8}$$

where * represents the excited state, and L and O_2 are the lumiphore and quencher, respectively.

The probability of the production of an oxygen singlet (excited state) is high and this is a strong indicator of the dynamic collision process. The collision quenching mechanism can be described by the Stern–Volmer equation [40]:

$$\frac{F_0}{F} = \frac{\tau_0}{\tau} = 1 + K_{SV}Q \tag{9.9}$$

where F_0 and τ_0 are the emitted luminescence intensity and decay time in the absence of the quencher respectively. Furthermore, F and τ are the same as previously but for a certain concentration of the quencher. Finally, Q is the concentration of the quencher and K_{SV} is the Stern–Volmer dynamic quenching constant given by [40]:

$$K_{SV} = k_q\tau_0 \tag{9.10}$$

where k_q is the bimolecular quenching constant. This holds given the intensity of emission is directly proportional to the luminophore's excited state population, assuming it does not change under continuous illumination as follows [35]:

$$\frac{dN}{dt} = f(t) - \gamma N_0 = 0 \tag{9.11}$$

$$\frac{dN}{dt} = f(t) - N(\gamma + k_q Q) = 0 \tag{9.12}$$

where N and N_0 are the excited state populations in the absence and presence of the quencher respectively, $f(t)$ defines the impinging optical power and γ is the radiative decay rate in the absence of the quencher. Combining (9.11) and (9.12) yields:

$$\frac{N_0}{N} = \frac{\gamma + k_q Q}{\gamma} \tag{9.13}$$

which clearly shows that the rate is inversely proportional to the lifetime, as will be shown in (9.15).

Phosphorescence quenching is a rate process of depopulation of the excited states and hence the following are valid [35]:

$$\tau_0 = \frac{1}{\gamma} \tag{9.14}$$

$$\tau = \frac{1}{\gamma + k_q Q} \tag{9.15}$$

and hence, the combination of (9.14) and (9.15) yields the Stern–Volmer equation as follows:

$$\frac{\tau_0}{\tau} = \frac{\gamma + k_q Q}{\gamma} = 1 + \frac{kqQ}{\tau} = 1 + k_q \tau_0 Q = 1 + k_{SV} Q \tag{9.16}$$

Due to the conservation of energy laws, the intensity is quenched as energy is removed from the system. The resulting variation in decay time can be attributed to the fact that during dynamic quenching there are multiple paths to non-radiative decay and therefore the rate at which the excited state is depopulated increases. In a state of equilibrium, in a solution of gas phase, the solubility of oxygen (sO_2) and the partial pressure of oxygen (pO_2) give the oxygen concentration (Q_{O_2}) [41]:

$$Q_{O_2} = sO_2 pO_2 \tag{9.17}$$

thus substituting into (9.16) yields:

$$\frac{\tau_0}{\tau} = 1 + K_{SV}^D Q_{O_2} = 1 + K_{SV}^P pO_2 \tag{9.18}$$

where K_{SV}^D and K_{SV}^P correspond to the Stern–Volmer coefficients for oxygen diffusion and permeability coefficients, respectively.

Until now, a linear constant gradient Stern–Volmer equation was assumed, see figure 9.3. However, most phosphorescent molecules used in oxygen sensing are embedded in host polymers for mechanical stability. In the vast majority of these cases, heterogeneity in the micro-environment results in deviations from this model. Myriad studies have focused on explaining the non-linearity of Stern–Volmer plots and, in most cases, it is attributed to a distribution of the phosphorescent probes on sites in a polymer medium that vary depending on the diffusion, permeability, solubility and lifetime characteristics [42, 43].

Specifically, the distribution of occupancy of sites depends on the attachment of luminophores on site specific parts of the thin layer of polymer but the quenching process rate distribution depends on the site specific oxygen permeability and/or oxygen diffusion coefficients on top of that [44]. It must be noted that this also means not observing a mono-exponential type of decay in temporal measurements thus multiple exponential models have also attempted to explain such phenomena but no unilateral solution has been given.

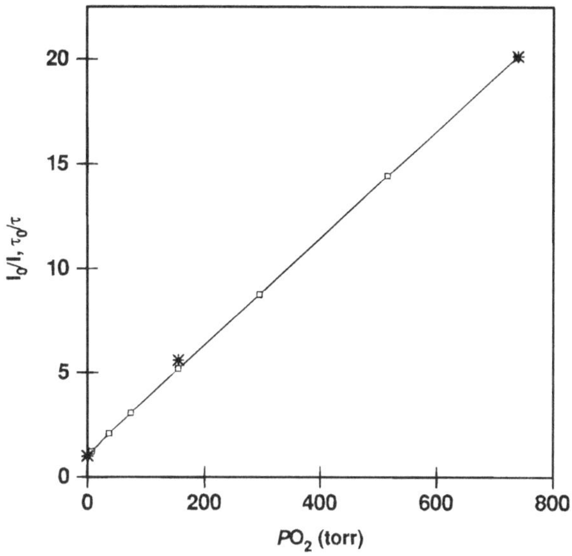

Figure 9.3. Stern–Volmer intensity (0) and weighted mean decay time plots (*) for PtOEPK in PS (at 30°C) versus applied oxygen partial pressure. Solid lines are best fits of a two-component model to the intensity quenching data [44]. Reprinted with permission from [44]. © (2020) American Chemical Society.

Focusing on the distribution based models, one of the most applied cases is the two-site model, which is the simplest case of the general idea of fractional contributions from each site with a different Stern–Volmer constant (multi-site model) [38]. It is obvious that a two-site kind of interpretation does not correspond to a plausible physical explanation of such a complex system, but has been proven to exhibit a high goodness of fit in various applications of oxygen dependent quenching of phosphorescence. The main principle of the multi-site theory follows [45]:

$$\frac{F_o}{F} = \frac{1}{\sum_{j=1} \frac{f_j}{1 + K_{SV}^j Q}} \tag{9.19}$$

where f_j is the fractional contribution of each site.

This approach is based on a random distribution model with arbitrary fractional weights. Thus, this corresponds to a multi-exponential model for the decay characteristics [42];

$$I(t) = \sum_{j=1} \left[\frac{f_{0j}}{\tau_{0j}} \right] e^{\frac{-t}{\tau_j}} \tag{9.20}$$

where $I(t)$ is the intensity over time of the received optical signal.

More physically plausible, is the log-Gaussian distribution in the lifetime and the bimolecular quenching rate. The choice of the specific parameters in the model considers that the two main kinetic processes involved in this type of device are

(i) the deactivation of the luminophore excited state, which is manifested mainly as changes in the lifetime in the absence of oxygen τ_0 and (ii) the quenching of the excited state by oxygen manifested as changes in the bimolecular constant k_q [46]. This model is based on the assumption of the following distribution equations:

$$\frac{\tau_{0,j}}{\tau_{0,\mathrm{mod}}} = \frac{k_{q,j}}{k_{q,\mathrm{mod}}} = e^{-x^2} \tag{9.21}$$

$$\rho_{1x} = \ln\left[\frac{\tau_{0,j}}{\tau_{0,\mathrm{mod}}}\right] \tag{9.22}$$

$$\rho_{2x} = \ln\left[\frac{k_{q,j}}{k_{q,\mathrm{mod}}}\right] \tag{9.23}$$

where the subscript mod indicates a modular value which is gained via the optimisation process.

To fit the data, optimal solutions are sought by fitting the observed data to different values of ρ and K_{SV}, usually for intensity measurements. Using this approach with experimental data typically reveals non-uniformities from the variation of k_q but not τ_0. Additionally, when the distribution widens, the sensitivity of the sensor is reduced. This has also led to partially successful predictions of lifetimes. By using this approach, different sensitivities manifested as complex decay kinetics have been given quantitative interpretation from simple distributions [46].

Simpler models suggest more empirical solutions, like adaptations of mathematical formulas from adsorption studies of heterogenous systems. Such are the Langmuir and the Freundlich isotherms [38, 40]. The Langmuir hypothesis states that only a fraction of the oxygen is adsorbed in monolayer phenomena [39, 47]:

$$\frac{Q_{ads}}{Q_{adsM}} = \frac{bp}{1 + bp} \tag{9.24}$$

where Q_{ads} and Q_{adsM} are the total and monolayer volumes of the adsorbate, respectively, b is the film thickness and recalling that p is the partial pressure.

The Freundlich isotherms correspond to gas-solid adsorption phenomena [47]:

$$\frac{Q_{ads}}{Q_{adsM}} = ap^{\frac{1}{n}} \tag{9.25}$$

Thus, their respective applications on the Stern–Volmer equation are as follows for Langmuir [48]:

$$\frac{F_0}{F} = 1 + \frac{K_{SV}Qbp}{1 + bp} \tag{9.26}$$

and for Freundlich [40]:

$$\frac{F_0}{F} = 1 + K_{SV} Q_{adsM} a p^{\frac{1}{n}} \tag{9.27}$$

However, approaches based on the last models, such as the non-linear solubility model that utilises the Langmuir model to explain the curved Stern–Volmer based on the non-linear solubility of oxygen, provided less satisfactory results compared to the two-site model [39]. Nevertheless, the simplicity of a Freundlich based model via the assignment of a power law has provided insights into the nature of the heterogeneity that led to non-linear Stern–Volmer plots. The most probable cause of these has been attributed to the distribution of the permeability of oxygen within the polymer-dye film [40].

9.3.3 Response and recovery time

A further critical parameter that must be considered is the response and recovery times of the oxygen sensor. This is important especially in dynamic systems where the goal is rapid monitoring of fast variations in the oxygen content of the environment. In some studies the assumption that the sensor film has constant thickness and the process is diffusion controlled are used to explain the differences observed experimentally in the gradients of concentration reductions near oxygen impermeable interfaces [49]. An empirical model found in the literature concerning the sensors' response time sensitivity can be found by fitting the Stern–Volmer equation as follows [50]:

$$t_{\rightarrow 90} = 3.06\left(\frac{b^2}{D}\right)\ln\left[10 + \frac{9K_{SV}\left(p\mathrm{O}_2^i - p\mathrm{O}_2^f\right)}{1 + p\mathrm{O}_2^i K_{SV}}\right] \tag{9.28}$$

where t is the response time after 90 s, D is the diffusion coefficient, b is the diffusive length and O_2^i and O_2^f are the initial and final oxygen levels.

It must be pointed out that the recovery time, or in other words the time it takes for the emitted luminescence in terms of the signal characteristics to follow the reduction of oxygen content in the environment under test, can vary significantly from the response time [51]. Generally, both response and recovery periods vary with phosphorescent dye, sensitivity of the indicator to changes in oxygen content, the polymer encapsulating material in terms of permeability and diffusion coefficients, thickness of the sensor as well as the oxygen content of the initial and final state [50].

9.4 Methods

Measurements to quantify the oxygen levels of a system that is in contact with a phosphor utilising its oxygen dependent quenching properties have led to three main ways of extracting information from the light signals; (i) intensity, (ii) phase and frequency (frequency domain) and (iii) lifetime (time domain) measurements.

9.4.1 Intensity

Steady state intensity measurements such as the ones used in pulse oximetry can be also used in the case of quenching of phosphorescence by taking advantage of the Stern–Volmer equation. Therefore, the intensity of emission and the oxygen concentration or partial pressure via Henry's law will be correlated. These measurements are effectively the sum of the de-excitations of electrons under constant illumination (i.e. under equilibrium). Nonetheless, they suffer from requiring additional calibration processes due to their dependence on geometry of the system in terms of distance and orientation of the light source and detector or in general temporal parameters that cause an additional instability and introduce error in the measurements [52]. Optical properties related to intensity measurements are also to a large extent dependent on the measurement apparatus, since differences in solid state devices such as their ability to transform electrons into photons and vice versa can vary largely. Moreover, the indicator and its concentration also introduce uncertainty, as a lack of uniformity can be embedded or encapsulated in the host [53].

One of the most prevalent issues with this type of method is also photobleaching of the luminescent dye used. Photobleaching is the process where the luminophores permanently lose their ability to emit light due to their interaction with their surroundings. In the case of phosphorescent indicators this phenomenon increases with oxygen concentration and is strongly related to the production of singlet excited oxygen states, with the goal of partially circumventing these limitations, referencing techniques are often employed [54].

9.4.2 Frequency domain lifetime

In this type of measurement the frequency and phase of the signal can be related to the oxygen concentration via the phase and frequency based apparent lifetimes calculation. This can be achieved by monitoring (i) the phase of the received signal with respect to the initial using a constant frequency, (ii) the demodulation factor as the frequency is varied and (iii) a combination of the two.

The variation of the frequency component of the signal passing through the medium under test leads to a lifetime dependent change in the demodulation factor [55]. A crucial note is that the frequency itself does not change. On the other hand, phase measurements take advantage of the phase change (i.e. time lag) that occurs, under continuous wave excitation, between the excitation and incident light. The correlation between the phase change and the lifetime of the phosphorescence allows time domain measurements of oxygen concentration by indirectly measuring lightwave properties of the emission using direct or cross-correlation phase detection systems [56]. This technique has had significant benefits for fluorescent measurements due to the short lifetimes of fluorescent phenomena and the difficulty of monitoring the decays in terms of equipment used. Nevertheless, it has also shown promising results in phosphorescent measurements of oxygen as well.

The equations of phase modulated fluorescent measurements stem from simple kinetic equations that are assumed to be true [35]. In detail, the optical signal transmitted $L(t)$ is a sinusoid that maintains its frequency as it transverses the medium:

$$L(t) = a + b \sin(\omega t) \tag{9.29}$$

where b/a is the modulation depth of the incident light. The excited state population $N(t)$ responds to the intensity variations with the insertion of a phase difference which appears as a 'time lag' at the receiver:

$$N(t) = A + B \sin(\omega t - \phi) \tag{9.30}$$

where ϕ is the phase induced. The intensity decay is then assumed to follow a single exponential after a hypothetical delta function point excitation. The time-dependent intensity of emission is given by the convolution integral of the illumination light with the impulse response function (which essentially is the mono-exponential decay we assumed to be true) [35]. This leads to an emission intensity governed by:

$$I(t) = I_0\tau[a + bm_\omega \cos(\omega t - \phi)] \tag{9.31}$$

where the induced phase is given by:

$$\tan(\phi) = \omega\tau_\phi \tag{9.32}$$

and τ_ϕ is the apparent phase-based lifetime. Furthermore, m_ω is the apparent demodulation factor for the specific frequency and is given as follows [57]:

$$m_\omega = \frac{1}{\sqrt{1 + \omega^2\tau_D^2}} \tag{9.33}$$

where τ_D is the apparent demodulation based lifetime given by [58]:

$$\tau_D = \frac{\sqrt{m_\omega^{-2} - 1}}{\omega} \tag{9.34}$$

It must be noted that this method does not provide the entire series of complex decay characteristics, as the multi-exponential and non-linear models are simplified to an apparent phase and demodulation correlated lifetime measurement. However, it does allow for precise and accurate determination of the oxygen quenching. In some cases the ratio of the two apparent measurements was used for this purpose.

This method was initially introduced as an alternative to time decay monitoring since it allows the correlation of measurements to the lifetime without directly measuring it. Instead it allows either detection of phase changes in the received signals or the observation of intensity changes whilst varying the sinusoid frequency. There is a direct analogy with the first order filtering of analogue electronic signals where either HPF or LPF responses can significantly dampen the amplitude of the input ac signals if they are above or below the cut-off frequency. A phase difference can also be introduced depending on the proximity of oscillation and the cut-off frequency. In the case of this method, the cut-off frequency is the frequency of excitation and the *RC* constant equates to the lifetime of the luminophore. The differential equations describing both cases are interchangeable.

An important method of determining the apparent lifetime is via the demodulation factor by alternating the modulation frequencies over a certain range and using (9.32). Simultaneously, the range of useful frequencies also gives a good indication of the modulation frequency applied for phase sensitive methods depending on the lifetime of the luminophore [59]. This is calculated by varying the frequency and monitoring the demodulation as well as the phase changes (see figure 9.4). It is different for every unique lifetime situation.

The optimal modulation point is situated where the phase changes are the most significant on variation of the lifetime. Such sensing platforms are usually calibrated *in situ* to give the best result according to the system at hand but a good indication can be inferred by plotting the graph. Theoretically, the optimal modulation frequency for phase measurements is $f_{opt} = 1/2\pi\tau$ (i.e. the angular frequency coincides exactly with the reciprocal of the lifetime) [60]. However, one must consider that the frequency changes within the experiments for oxygen quenching measurements where the oxygen concentration is proportional to the lifetime. Also, when considering the errors introduced from the detectors this has to be divided by $\sqrt{2}$ [60].

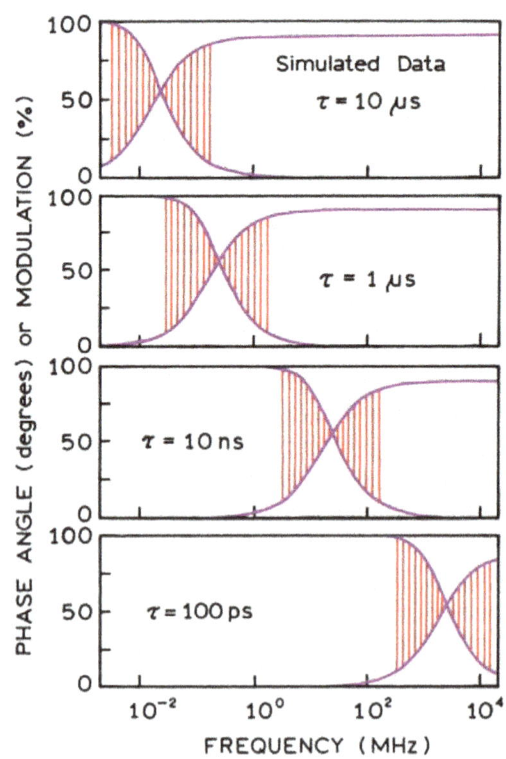

Figure 9.4. Frequency *vs* phase angle for different lifetimes [35]. Reprinted with permission from Springer: [35] © (2013).

A common approach to the determination of the fluorimetric phase changes is cross-correlation detection [35]. The main functionality of this type of measurement is the transformation of the high frequency signal to a low frequency one whilst preserving both the phase and demodulation information simultaneously. The benefits that stem from this are prevalent in fluorescent spectroscopy where the modulation frequencies were high and difficult to process [61]. This is achieved by transmission and reception of a reference signal that maintains its frequency characteristic over the medium as before. The reference and received signals are firstly multiplied by an expected, scaled version of themselves as follows [56]:

$$I_R(t) = I_{R0}\left[1 + m_R \cos\left(\omega t - \phi_R\right)\right] \tag{9.35}$$

$$I_T(t) = I_{T0}\left[1 + m_T \cos\left(\omega t - \phi_T\right)\right] \tag{9.36}$$

where the T and R subscripts indicate the transmitted and received signals, respectively, and m is an amplitude scaling factor.

The next step is to filter to remove the unwanted high frequency component, which leaves just the phase and demodulation information [56]:

$$\int I_T(t)I_R(t)\mathrm{d}t = \frac{(I_{R0}m_R)(I_{T0}m_T)}{2}[\cos(\Delta\omega t + \Delta\phi)] \tag{9.37}$$

where Δ_ω is the angular frequency difference of the two signals and Δ_ϕ is the phase difference between the two signals. Equation (9.37) does not oscillate in time or slowly drift since the frequency difference between the signals is preferably designed to be small. Thus, by controlling the properties of the reference signal it is possible to get information about the phase and the demodulation factor.

9.4.3 Time domain lifetime

Direct lifetime measurements rely on the analysis of the luminescent emission after pulsed excitation by focusing on the intensity of emission over time in the decay profiles [62]. In order to achieve this, numerical based logarithmic procedures are used by fitting parts of or the entire decay into exponential models and subsequently extracting the lifetime based on least squares regression (linear or non-linear). In simpler cases this involves the monitoring of the time it takes to reach a checkpoint intensity or of the areas under the curve in the decay profiles.

The lifetime is then inserted to the Stern–Volmer equation, which in turn correlates it to the oxygen concentration as well as a multitude of other parameters such as the indicator material and its surrounding environment. It must be noted that due to the emphasis on post-processing methods after a decay profile has been captured these types of measurements are simpler to implement with robust solid state electronics.

This approach allows for information extraction that is to a large extent not dependent on luminescent dye concentration, macro-scale spatial non-uniformities of the excitation light source and detector, and the macro-scale uniformity of the distribution of the indicator onto the support medium [63]. One important aspect of this type of approach is that direct lifetime determination through pulse decays is much simpler than the corresponding frequency domain measurements in terms of instrumentation and complexity of experiments. The reason for this is that they do not require reference points as well as optical wavelength filters, since the light source is switched in a pulsed configuration. An additional benefit is that the production of such pulsed excitations is easy to generate and miniaturise [10]. It must be noted that the lifetime of the indicator should be known *a priori* due to the fact that the instrumentation must be operational and optimised over the time decays involved.

In order to arrive at a lifetime from a luminescent decay a wide variety of models have been adopted to explain the decay profile observed in the optical sensors. Primarily, the exponential models are preferred due to their good fitness and connection with the underlying physical mechanisms used to explain phosphorescence [38, 64, 65]. Single exponential models seem to be a good fit in the absence of the quencher, i.e. oxygen in this case, for most dyes and their encapsulation media deviate for higher concentrations. To attribute a lifetime in such cases the curve is fitted to a multiple (with the simplest case being the double) or 'stretched' exponential [66].

Moreover, in some cases an additional term was added to compensate for possible shifts in the y-axis due to external factors such as ambient light as follows [66]:

$$I_{\mathrm{de}}(t) = a\exp\left[\frac{-t}{\tau_a}\right] + b\exp\left[\frac{-t}{\tau_b}\right] + c \tag{9.38}$$

while in general terms, the 'stretched' exponential follows [66]:

$$I_{\mathrm{se}}(t) = a\exp\left[-\left(\frac{t}{\tau_a}\right)^{\beta}\right] + c \tag{9.39}$$

recalling that $I(t)$ is the decay as a function of the intensity of the signal over time, refer to figure 9.5. The subscripts 'se' and 'de' denote the double and stretched exponential models, respectively, and a and b are the fractional contributions from each exponential decay. The terms τ_a and τ_b are the respective pseudo lifetimes, β is the power law coefficient of the stretched exponential model and c the coefficient to compensate for the displacement away from the y-axis.

In order to arrive at an average lifetime for the double exponential model (9.38), both contributions must be taken into account, yielding:

$$\langle\tau_{\mathrm{de}}\rangle = \frac{a\tau_a + b\tau_b}{a + b} \tag{9.40}$$

For the double exponential model, the existence of two distinct lifetimes reflects two different sites that each give a different fractional contribution towards the decay. For the stretched exponential it is assumed that the decay originates from the sites-lifetimes distribution $\rho(\tau)$, given by [67]:

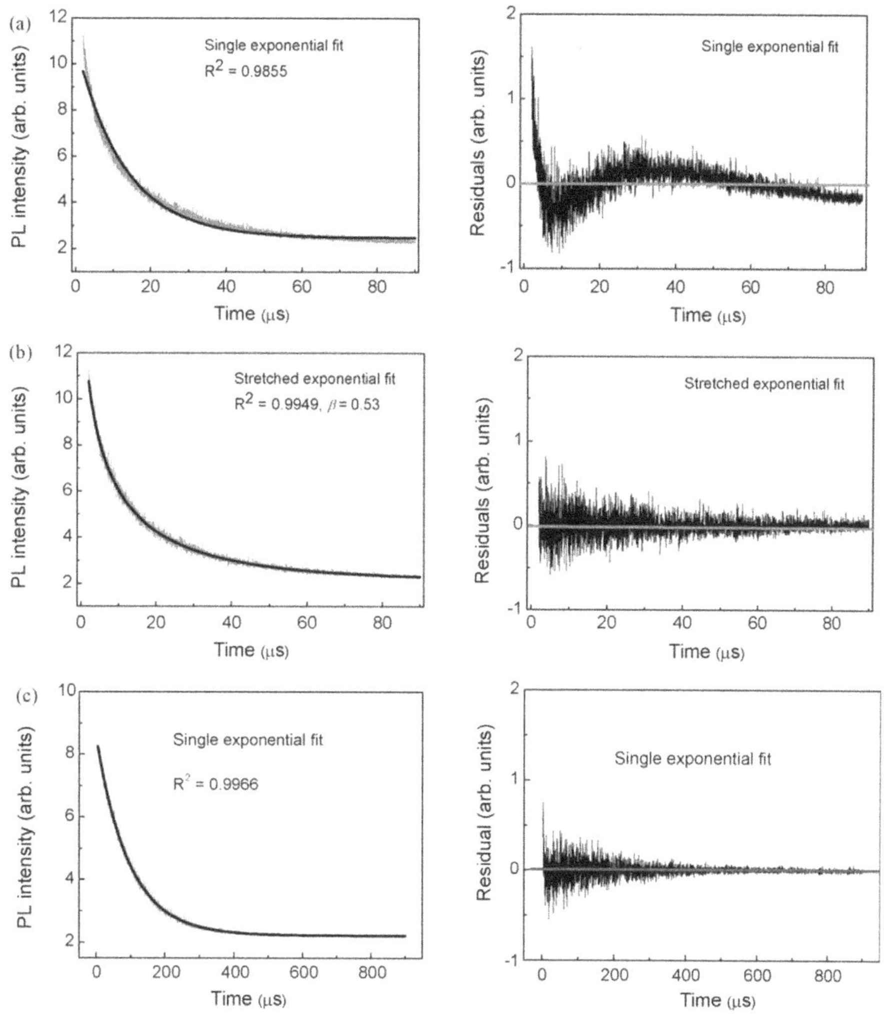

Figure 9.5. Experimental data of the decays following a pulse LED excitation along with the fitted curves of a single exponential and a stretched decay model: (a), (b) for an oxygen level of 40 ppm and the single exponential fit, (c) for a decay in a nitrogen or argon environment. The residuals of each fitting method are also given [67]. Reprinted from [67]. © (2010), with permission from Elsevier.

$$I_{se}(t) = \int_0^{\infty} \exp\left[-\frac{t}{\tau}\right]\rho(\tau)\ \mathrm{d}\tau \tag{9.41}$$

To give the lifetime for the stretched exponential function, the ensemble average is used [66]:

$$\langle \tau_{se} \rangle = \frac{\tau_a}{\beta}\Gamma\left(\frac{1}{\beta}\right) \tag{9.42}$$

where Γ(.) is the gamma function [66].

With this method the standard deviation and the width of the distribution can also be calculated:

$$\sigma^{1/2} = \langle\tau_{se}\rangle w = \langle\tau_{se}\rangle\frac{\beta\Gamma\left(2\beta^{-1}\right) - \Gamma^2\left(\beta^{-1}\right)}{\Gamma\left(\beta^{-1}\right)} \quad (9.43)$$

where σ is the standard deviation and w is the width of the distribution

The values of *sigma* and the divergence of β from one (single exponential decay case) can be correlated to the micro-heterogeneity that gives rise to a non-linear Stern–Volmer plot.

However, these types of measurements require non-linear fitting methods that tend to be more complex and time consuming upon their implementation. Alternatively, models such as the phase plane method use linear processes to analyse mono-exponential models by exploiting the relation between the integrated decay profile with the original one in order to determine the lifetime from the slope of (9.44) [68, 69]:

$$Y(t) = -\frac{1}{\tau}I(t) + C \quad (9.44)$$

where $Y(t)$ is the original mono-exponential decay profile over a certain window of time, $I(t)$ is the integrated values of the same window time and C is a constant. It must be noted that for double exponential models additional steps are required for its application.

Furthermore, arguably the simplest case of deriving the lifetime from a decay profile, the rapid lifetime determination algorithm, does not require any form of fitting but rather relies on the integrated values from a multitude of time windows from the decay curve [70]:

$$\tau = \frac{\Delta t}{\ln\left(\frac{D_1}{D_2}\right)} \quad (9.45)$$

where Δt is the time window length and D_1 and D_2 are the integrated values of the corresponding window.

9.5 Instrumentation

At this point the equipment and experimental setups that have been applied using the methods previously described are reviewed. It must be noted that this section is divided into two main categories, single point and imaging techniques. Basic arrays of photodiodes will be considered in the first category due to their limited resolution and greater similarities with the first class of devices.

9.5.1 Single point sensing

In the case of optical haemoglobin-based ratiometric sensors, the main components of the system consist of solid state electronics for the light source and detector, i.e. LED and photodiode, as well as for the processing and transmission of the electrical signals produced. As explained in the previous section, the principles of operation of such systems rely on dual wavelength measurements, thus in most cases IR and red LEDs are preferred due to better tissue penetration and reduced scattering effects [71].

In some cases, arrays of OLEDs and organic photodiodes have been used [31] with the choice of organic based semiconductor devices seeming preferable due to their mechanical properties and specifically that of high flexibility and stretchability [72] (see figure 9.6 for an illustrative example). Moreover, planar optodes have been extensively used with various phosphorescent based oxygen measurements. Systems involving simpler time domain measurements rely on pulse excitation of the light source (usually LED) and the use of a digital oscilloscope after the photodiode to observe the decay, although detection is sometimes performed by avalanche photodetectors in order to achieve a higher intensity electronic signal [73].

Fascinatingly enough, such optode based detectors that take advantage of luminescent indicator optical properties have been miniaturised with instrumentation that even utilised frequency domain time resolved measurements [74]. Newer approaches in this domain involved basic analogue and digital processing solutions that led to simple and accurate phase measurements [55, 56]. Later studies have explored the use of all three main ways of extracting the quenching effect of oxygen as described in the previous section. Interestingly, phase resolved phosphorimetry has been extensively studied with novel implementations for high accuracy results [75]. This has also led to dual luminophore implementation for a two-wavelength referencing system [76].

A wide topic of research has been the implementation of oxygen sensing via fibre and more generally plasmonic optics. The first implementations for monitoring

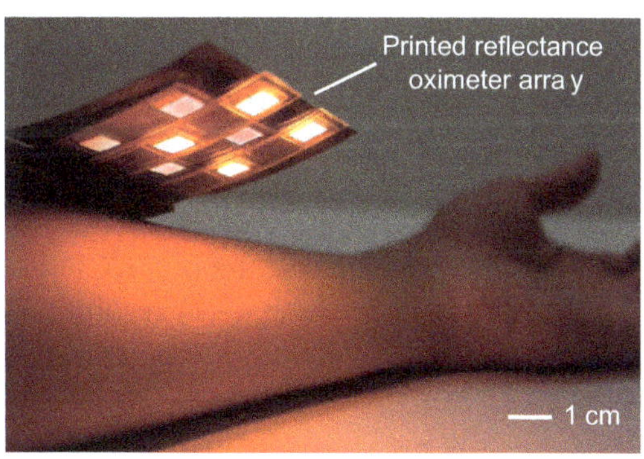

Figure 9.6. Device for spatially resolved tissue oxygen saturation levels based on an array of OLEDs and OPDs [31]. Reproduced from [31]. CC BY 4.0.

oxygen partial pressure involved using phosphorescent indicator dye encapsulation onto the fibre optic tip material and led to *in vivo* studies of oxygen partial pressure in the blood stream [77, 78]. The oxygen dependent quenching phosphor should be embedded in or placed prior to an oxygen permeable glass substrate. For higher sensitivities an optically insulating layer is desirable [79]. In this direction multimode fibres have been implemented to monitor both intra- and inter-cellular oxygen concentration [80]. Additionally, placing the indicators on the cladding layers of the fibre allowed for optical fibre evanescent wave based detection [81, 82].

Interestingly, other plasmonic techniques have focused on analyte detection without the use of optical fibres, such as waveguide/lightguide systems [83]. Efforts have led to multimode waveguides with a multitude of embedded fluorophores and phosphors for simultaneous multi-analyte detection [84]. These also take advantage of the use of simple solid state detector systems and light sources as well as low cost manufacturing.

Simple micro-controller processing units have been used both for capturing and processing the indicator luminescence [9, 10] as well as wireless implementations of pulse oximeters [85]. This allows for the development of simple, cheap and robust devices for single point monitoring of a single or multiple photodiodes. There are a wide variety of micro-controllers with ADCs that have 12-bit resolution and numerous ports to allow for arrays of photodetectors to be used in future implementations.

For the transduction of the light detector signal, transimpedance amplifiers have generally been used with the goal of transforming the current signal, from the photodiodes, into voltage and at the same time enhance it. To power and modulate the light sources in the case of LEDs, function generators were sometimes used [86].

It must be pointed out that optical as well as analogue electrical filtering has been extensively used throughout luminescent oxygen sensor development. Their purpose has focused on filtering the carrier (optical filters) and modulation frequency spectra of the signal towards achieving a better signal-to-noise ratio (SNR) by blocking the frequencies that were unnecessary [87], i.e. source of error. In some cases RC filters have been used for analogue integration and multipliers for the extraction of the phase characteristics of the light signal [56]. Along this direction, comparators have been used as a part of analogue signal processing equipment to get a direct sense of the phase shift in terms of time delay [74] instead of intensity differences.

9.5.2 Imaging techniques

The most common imaging setup involves the use of a CCD or CMOS camera with the goal of monitoring the emitted light with a 2D resolution. A thin luminescent film is usually preferred as the indicator of oxygen content. Optical filters are in most cases employed for either excitation or emission wavelength selection, or both. This method requires in most cases the use of a computer to log the data and manage the processing [88]. Lifetime measurements require time gated CCD cameras, whose basic principle of operation involves turning the camera on and off, while being synchronised to start the measurements after the decay of the luminophores has started, and then integrate the intensity of illumination over the time to acquire the integrals and then the lifetime is calculated via equation (9.46) [21]. Advancements in this field have led to miniaturised cameras that have even been applied to *in vivo* studies [89].

$$r = \frac{t_2 - t_1}{\ln\left(\frac{A_1}{A_2}\right)} \tag{9.46}$$

where the fraction r gives a sense of the lifetime of the decay and A_1, A_2 are the areas under the decay curve for the two identical sets of time intervals starting at t_1, t_2.

Fibre optic solutions usually provide single point measurements but there have been novel devices developed where spatially resolved measurements were applied by simultaneously relying on multiple bundles of fibre. Particularly, pulsed excitation measurements were used in combination with a stepper motor to move through a 2D area [90] or a fibre plasmonic array [91]. Impressively, some fluorescent based sensors have appeared to utilise embedding of the luminophore in the cladding layers so that the evanescent wave properties can provide spatially resolved read outs [92].

More accurate optical methods have been deployed with the use of two-photon imaging techniques resulting in a spatial resolution of sub-millimetre accuracy in three dimensions [93]. This involved the use of a two-photon microscope that utilises excitation laser sources and high accuracy scanning platforms along with optical filters. A good example of implementation of such equipment in studies was the use of two-photon phosphorescent lifetime microscopy for the monitoring of blood flow and oxygenation in small blood vessels of rats [93].

Fluorescent lifetime imaging (FLIM) studies involved the use of a modified time correlated single photon counting (TCSPC) platform. This achieved the imaging of cellular oxygen consumption in micrometre resolution [94]. This was then used for phosphorescence lifetime measurements (PLIM), in combination with FLIM techniques for referencing purposes, and thus oxygen dependent quenching was monitored with accurate predictions in the longer time-scale of phosphorescence [95, 96]. It must be noted that conventional TCSPC equipment is not suitable for phosphorescent measurements but needs to be subjected to the appropriate modifications. The applications range from *in vivo* studies to monitoring microfluidic bioreactors and cell cultures [97].

In this direction multi photon microscopy is applied to achieve even greater sub micron resolution. This requires frequency-multiplexing (parallel excitation–detection) and a line scanning lens microscope system [98]. It must be noted that this is a high complexity methodology.

9.6 Summary

In summary, this chapter gave a detailed overview of the main applications of VLC in health monitoring by focusing on oxygen sensing. The examples of applications provided here involved cancer studies and cell culture monitoring. Furthermore, the theoretical background and the principles of operation of oxygen detection were given for the cases of pulse oximetry and quenching of phosphorescence via intensity, frequency, as well as direct lifetime measurements. These mechanisms, in the authors' collective opinion, are the most prominent candidates for future applications and specifically direct lifetime determination due to the ease of application with modern electronics and the capability of miniaturisation. In order to complete the picture, the principal instrumentation followed in the literature was presented with the two focal points being single sensing and imaging techniques along with a description of the technology utilised. It must be underlined that only optoelectronic solutions are described, but it is the authors' opinion that the advantages of this field compared to electrochemical solutions are significant. To conclude, advances in this domain have the potential to lead to a new era of medical devices that will bring forth the personalisation of healthcare and constitute an irreplaceable tool for medical practitioners.

References

[1] Lochner C M, Khan Y, Pierre A and Arias A C 2014 All-organic optoelectronic sensor for pulse oximetry *Nat. Commun.* **5** 5745

[2] Myers F B and Lee L P 2008 Innovations in optical microfluidic technologies for point-of-care diagnostics *Lab Chip* **8** 2015–31

[3] Büchele P *et al* 2015 X-ray imaging with scintillator-sensitized hybrid organic photodetectors *Nat. Photonics* **9** 843–8

[4] Kong L, Zhao Y, Dong L, Jian Y, Jin X, Li B, Feng Y, Liu M, Liu X and Wu H 2013 Non-contact detection of oxygen saturation based on visible light imaging device using ambient light *Opt. Express* **21** 17464–71

[5] Kim J *et al* 2017 Miniaturized battery-free wireless systems for wearable pulse oximetry *Adv. Funct. Mater.* **27** 1604373

[6] Rachim V P and Chung W-Y 2019 Wearable-band type visible-near infrared optical biosensor for non-invasive blood glucose monitoring *Sensors Actuators* B **286** 173–80

[7] Zherebtsov E, Doronin A, Bykov A, Popov A and Meglinski I 2017 Impact of blood volume on the diffuse reflectance spectra of human skin in visible and NIR spectral ranges *Proc. SPIE* **10412** 104120H

[8] Kwok N-Y, Dong S, Lo W and Wong K-Y 2005 An optical biosensor for multi-sample determination of biochemical oxygen demand (BOD) *Sensors Actuators* B **110** 289–98

[9] Mathupala S P, Kiousis S and Szerlip N J 2016 A lab assembled microcontroller-based sensor module for continuous oxygen measurement in portable hypoxia chambers *PLoS One* **11** e0148923

[10] Palma A J, López-González J, Asensio L J, Fernández-Ramos M D and Capitán-Vallvey L F 2007 Microcontroller-based portable instrument for stabilised optical oxygen sensor *Sensors Actuators* B **121** 629–38

[11] Eltzschig H K and Carmeliet P 2011 Hypoxia and inflammation *N. Engl. J. Med.* **364** 656–65

[12] Wilson W R and Hay M P 2011 Targeting hypoxia in cancer therapy *Nat. Rev. Cancer* **11** 393–410

[13] Graeber T G, Osmanian C, Jacks T, Housman D E, Koch C J, Lowe S W and Giaccia A J 1996 Hypoxia-mediated selection of cells with diminished apoptotic potential in solid tumours *Nature* **379** 88–91
[14] Vaupel P and Mayer A 2007 Hypoxia in cancer: significance and impact on clinical outcome *Cancer Metastasis Rev.* **26** 225–39
[15] Vinogradov S A, Lo L-W, Jenkins W T, Evans S M, Koch C and Wilson D F 1996 Noninvasive imaging of the distribution in oxygen in tissue *in vivo* using near-infrared phosphors *Biophys. J.* **70** 1609–17
[16] Vaupel P, Fortmeyer H P, Runkel S and Kallinowski F 1987 Blood flow, oxygen consumption, and tissue oxygenation of human breast cancer xenografts in nude rats *Cancer Res.* **47** 3496–503
[17] Harris A L 2002 Hypoxia-a key regulatory factor in tumour growth *Nat. Rev. Cancer* **2** 38–47
[18] Zhang S, Hosaka M, Yoshihara T, Negishi K, Iida Y, Tobita S and Takeuchi T 2010 Phosphorescent light-emitting iridium complexes serve as a hypoxia-sensing probe for tumor imaging in living animals *Cancer Res.* **70** 4490–8
[19] Hodges S J and Atala A 2014 Tissue-engineered organs *Principles of Tissue Engineering* (New York: Elsevier) pp 1765–77
[20] Hussein K H, Park K-M, Kang K-S and Woo H-M 2016 Biocompatibility evaluation of tissue-engineered decellularized scaffolds for biomedical application *Mater. Sci. Eng.* C **67** 766–78
[21] Kellner K, Liebsch G, Klimant I, Wolfbeis O S, Blunk T, Schulz M B and Göpferich A 2002 Determination of oxygen gradients in engineered tissue using a fluorescent sensor *Biotechnol. Bioeng.* **80** 73–83
[22] Mehta G, Mehta K, Sud D, Song J W, Bersano-Begey T, Futai N, Heo Y S, Mycek M-A, Linderman J J and Takayama S 2007 Quantitative measurement and control of oxygen levels in microfluidic poly (dimethylsiloxane) bioreactors during cell culture *Biomed. Microdevices* **9** 123–34
[23] Malda J, Rouwkema J, Martens D E, le Comte E P, Kooy F K, Tramper J, van Blitterswijk C A and Riesle J 2004 Oxygen gradients in tissue-engineered PEGT/PBT cartilaginous constructs: measurement and modeling *Biotechnol. Bioeng.* **86** 9–18
[24] Clark L C Jr, Wolf R, Granger D and Taylor Z 1953 Continuous recording of blood oxygen tensions by polarography *J. Appl. Physiol.* **6** 189–93
[25] Wang X-D and Wolfbeis O S 2014 Optical methods for sensing and imaging oxygen: materials, spectroscopies and applications *Chem. Soc. Rev.* **43** 3666–761
[26] Sudhakar M, Djurovich P I, Hogen-Esch T E and Thompson M E 2003 Phosphorescence quenching by conjugated polymers *J. Am. Chem. Soc.* **125** 7796–7
[27] Campani G, Ribeiro M P A, Horta A C L, Giordano R C, Badino A C and Zangirolami T C 2015 Oxygen transfer in a pressurized airlift bioreactor *Bioprocess Biosyst. Eng.* **38** 1559–67
[28] Swinehart D F 1962 The Beer–Lambert law *J. Chem. Educ.* **39** 333
[29] Parnis J M and Oldham K B 2013 Beyond the Beer–Lambert law: the dependence of absorbance on time in photochemistry *J. Photochem. Photobiol.* A **267** 6–10
[30] Tremper K K and Barker S J 1989 Pulse oximetry *Anesthesiol.: J. Am. Soc. Anesthesiol.* **70** 98–108
[31] Khan Y *et al* 2018 A flexible organic reflectance oximeter array *Proc. Natl Acad. Sci. USA* **115** E11015–24

[32] Khan Y, Ostfeld A E, Lochner C M, Pierre A and Arias A C 2016 Monitoring of vital signs with flexible and wearable medical devices *Adv. Mater.* **28** 4373–95
[33] Chan E D, Chan M M and Chan M M 2013 Pulse oximetry: understanding its basic principles facilitates appreciation of its limitations *Respir. Med.* **107** 789–99
[34] Sinex J E 1999 Pulse oximetry: principles and limitations *Am. J Emerg. Med.* **17** 59–66
[35] Lakowicz J R 2013 *Principles of Fluorescence Spectroscopy* (Berlin: Springer)
[36] Quaranta M, Borisov S M and Klimant I 2012 Indicators for optical oxygen sensors *Bioanal. Rev.* **4** 115–57
[37] Gewehr P M and Delpy D T 1993 Optical oxygen sensor based on phosphorescence lifetime quenching and employing a polymer immobilised metalloporphyrin probe *Med. Biol. Eng. Comput.* **31** 11–21
[38] Carraway E R, Demas J N and DeGraff B A 1991 Photophysics and oxygen quenching of transition-metal complexes on fumed silica *Langmuir* **7** 2991–8
[39] Demas J N, DeGraff B A and Xu W 1995 Modeling of luminescence quenching-based sensors: comparison of multisite and nonlinear gas solubility models *Anal. Chem.* **67** 1377–80
[40] Douglas P and Eaton K 2002 Response characteristics of thin film oxygen sensors, Pt and Pd octaethylporphyrins in polymer films *Sensors Actuators* B **82** 200–8
[41] Forstner H and Gnaiger E 1983 Calculation of equilibrium oxygen concentration *Polarographic Oxygen Sensors* (Berlin: Springer) pp 321–33
[42] Albery W J, Bartlett P N, Wilde C P and Darwent J R 1985 A general model for dispersed kinetics in heterogeneous systems *J. Am. Chem. Soc.* **107** 1854–8
[43] Mills A 1997 Optical oxygen sensors *Platin. Met. Rev.* **41** 115–27
[44] Hartmann P and Trettnak W 1996 Effects of polymer matrices on calibration functions of luminescent oxygen sensors based on porphyrin ketone complexes *Anal. Chem.* **68** 2615–20
[45] Parker C A 1968 *Photoluminescence of Solutions: With Applications to Photochemistry and Analytical Chemistry* (New York: Elsevier)
[46] Mills A 1999 Response characteristics of optical sensors for oxygen: a model based on a distribution in τ_o and k_q *Analyst* **124** 1309–14
[47] Beswick R B and Pitt C W 1988 Oxygen detection using phosphorescent Langmuir–Blodgett films of a metalloporphyrin *Chem. Phys. Lett.* **143** 589–92
[48] Charlesworth J M and Gan T H 1997 Quenching of ketone phosphorescence by molecular oxygen at the gas/solid interface *Langmuir* **13** 2699–707
[49] Mills A and Chang Q 1992 Modelled diffusion-controlled response and recovery behaviour of a naked optical film sensor with a hyperbolic-type response to analyte concentration *Analyst* **117** 1461–6
[50] Mills A and Lepre A 1997 Controlling the response characteristics of luminescent porphyrin plastic film sensors for oxygen *Anal. Chem.* **69** 4653–9
[51] Xue R, Behera P, Xu J, Viapiano M S and Lannutti J J 2014 Polydimethylsiloxane core–polycaprolactone shell nanofibers as biocompatible, real-time oxygen sensors *Sensors Actuators* B **192** 697–707
[52] Papkovsky D B 2004 Methods in optical oxygen sensing: protocols and critical analyses *Methods in Enzymology* vol 381 (New York: Elsevier) pp 715–35
[53] Carraway E R, Demas J N and DeGraff B A 1991 Luminescence quenching mechanism for microheterogeneous systems *Anal. Chem.* **63** 332–6
[54] Hartmann P, Leiner M J P and Kohlbacher P 1998 Photobleaching of a ruthenium complex in polymers used for oxygen optodes and its inhibition by singlet oxygen quenchers *Sensors Actuators* B **51** 196–202

[55] Wang F, Chang J, Chen X, Wang Z, Wang Q, Wei Y and Qin Z 2017 Optical fiber oxygen sensor utilizing a robust phase demodulator *Measurement* **95** 1–7

[56] Chu C-S, Lin K-Z and Tang Y-H 2016 A new optical sensor for sensing oxygen based on phase shift detection *Sensors Actuators* B **223** 606–12

[57] McGown L B and Bright F V 1984 Phase-resolved fluorescence spectroscopy *Anal. Chem.* **56** 1400A–17A

[58] Diamandis E P 1988 Immunoassays with time-resolved fluorescence spectroscopy: principles and applications *Clin. Biochem.* **21** 139–50

[59] Szmacinski H and Lakowicz J R 1996 Frequency-domain lifetime measurements and sensing in highly scattering media *Sensors Actuators* B **30** 207–15

[60] Ogurtsov V I and Papkovsky D B 1998 Selection of modulation frequency of excitation for luminescence lifetime-based oxygen sensors *Sensors Actuators* B **51** 377–81

[61] Lakowicz J R and Berndt K W 1991 Lifetime-selective fluorescence imaging using an RF phase-sensitive camera *Rev. Sci. Instrum.* **62** 1727–34

[62] Gratton E, Breusegem S, Sutin J D B, Ruan Q and Barry N P 2003 Fluorescence lifetime imaging for the two-photon microscope: time-domain and frequency-domain methods *J. Biomed. Opt.* **8** 381–91

[63] Draxler S, Lippitsch M E, Klimant I, Kraus H and Wolfbeis O S 1995 Effects of polymer matrixes on the time-resolved luminescence of a ruthenium complex quenched by oxygen *J. Phys. Chem.* **99** 3162–7

[64] Chen X, Zhong Z, Li Z, Jiang Y, Wang X and Wong K 2002 Characterization of ormosil film for dissolved oxygen-sensing *Sensors Actuators* B **87** 233–8

[65] Eaton K, Douglas B and Douglas P 2004 Luminescent oxygen sensors: time-resolved studies and modelling of heterogeneous oxygen quenching of luminescence emission from Pt and Pd octaethylporphyrins in thin polymer films *Sensors Actuators* B **97** 2–12

[66] Lee K C B, Siegel J, Webb S E D, Leveque-Fort S, Cole M J, Jones R, Dowling K, Lever M J and French P M W 2001 Application of the stretched exponential function to fluorescence lifetime imaging *Biophys. J.* **81** 1265–74

[67] Cai Y, Smith A, Shinar J and Shinar R 2010 Data analysis and aging in phosphorescent oxygen-based sensors *Sensors Actuators* B **146** 14–22

[68] Collier B B and McShane M J 2013 Time-resolved measurements of luminescence *J. Lumin.* **144** 180–90

[69] Adamson A W and Demas J N 1971 Evaluation of photoluminescence lifetimes *J. Phys. Chem.* **75** 2463–6

[70] Woods R J, Scypinski S and Love L J C 1984 Transient digitizer for the determination of microsecond luminescence lifetimes *Anal. Chem.* **56** 1395–400

[71] Andresen V, Alexander S, Heupel W-M, Hirschberg M, Hoffman R M and Friedl P 2009 Infrared multiphoton microscopy: subcellular-resolved deep tissue imaging *Curr. Opin. Biotechnol.* **20** 54–62

[72] Root S E, Savagatrup S, Printz A D, Rodriquez D and Lipomi D J 2017 Mechanical properties of organic semiconductors for stretchable, highly flexible, and mechanically robust electronics *Chem. Rev.* **117** 6467–99

[73] Ji S *et al* 2009 Real-time monitoring of luminescent lifetime changes of PtOEP oxygen sensing film with LED/photodiode-based time-domain lifetime device *Analyst* **134** 958–65

[74] Trettnak W, Kolle C, Reininger F, Dolezal C and O'leary P 1996 Miniaturized luminescence lifetime-based oxygen sensor instrumentation utilizing a phase modulation technique *Sensors Actuators* B **36** 506–12

[75] Chatni M R, Li G and Porterfield D M 2009 Frequency-domain fluorescence lifetime optrode system design and instrumentation without a concurrent reference light-emitting diode *Appl. Opt.* **48** 5528–36

[76] Valledor M, Campo J C, Sánchez-Barragán I, Viera J C, Costa-Fernández J M and Sanz-Medel A 2006 Luminescent ratiometric method in the frequency domain with dual phase-shift measurements: application to oxygen sensing *Sensors Actuators* B **117** 266–73

[77] Peterson J I, Fitzgerald R V and Buckhold D K 1984 Fiber-optic probe for *in vivo* measurement of oxygen partial pressure *Anal. Chem.* **56** 62–7

[78] Papkovsky D B 1995 New oxygen sensors and their application to biosensing *Sensors Actuators* B **29** 213–8

[79] Klimant I, Meyer V and Kühl M 1995 Fiber-optic oxygen microsensors, a new tool in aquatic biology *Limnol. Oceanogr.* **40** 1159–65

[80] Park E J, Reid K R, Tang W, Kennedy R T and Kopelman R 2005 Ratiometric fiber optic sensors for the detection of inter-and intra-cellular dissolved oxygen *J. Mater. Chem.* **15** 2913–9

[81] MacCraith B D, McDonagh C M, O'Keeffe G, Keyes E T, Vos J G, O'Kelly B and McGilp J F 1993 Fibre optic oxygen sensor based on fluorescence quenching of evanescent-wave excited ruthenium complexes in sol–gel derived porous coatings *Analyst* **118** 385–8

[82] Singer E, Duveneck G L, Ehrat M and Widmer H M 1994 Fiber optic sensor for oxygen determination in liquids *Sensors Actuators* A **42** 542–6

[83] Lee K S, Lee H L T and Ram R J 2007 Polymer waveguide backplanes for optical sensor interfaces in microfluidics *Lab Chip* **7** 1539–45

[84] Burke C S, McGaughey O, Sabattié J-M, Barry H, McEvoy A K, McDonagh C and MacCraith B D 2005 Development of an integrated optic oxygen sensor using a novel, generic platform *Analyst* **130** 41–5

[85] Timm U, McGrath D, Lewis E, Kraitl J and Ewald H 2009 Sensor system for non-invasive optical hemoglobin determination *Sensors, 2009 IEEE* pp 1975–8

[86] Bambot S B, Rao G, Romauld M, Carter G M, Sipior J, Terpetchnig E and Lakowicz J R 1995 Sensing oxygen through skin using a red diode laser and fluorescence lifetimes *Biosens. Bioelectron.* **10** 643–52

[87] Shen L, Ratterman M, Klotzkin D and Papautsky I 2011 A CMOS optical detection system for point-of-use luminescent oxygen sensing *Sensors Actuators* B **155** 430–5

[88] Hartmann P, Ziegler W, Holst G and Lübbers D W 1997 Oxygen flux fluorescence lifetime imaging *Sensors Actuators* B **38** 110–5

[89] Rector D M, Poe G R, Redgrave P and Harper R M 1997 A miniature CCD video camera for high-sensitivity light measurements in freely behaving animals *J. Neurosci. Methods* **78** 85–91

[90] Chaturvedi P, Hauser B A, Foster J S, Karplus E, Levine L H, Coutts J L, Richards J T, Vanegas D C and McLamore E S 2014 A multiplexing fiber optic microsensor system for monitoring spatially resolved oxygen patterns *Sensors Actuators* B **196** 71–9

[91] Rigo M V and Geissinger P 2012 Crossed optical fiber sensor arrays for high-spatial-resolution sensing: application to dissolved oxygen concentration measurements *J. Sens.* **2012**

[92] Prince B J, Schwabacher A W and Geissinger P 2001 A readout scheme providing high spatial resolution for distributed fluorescent sensors on optical fibers *Anal. Chem.* **73** 1007–15

[93] Lecoq J, Parpaleix A, Roussakis E, Ducros M, Houssen Y G, Vinogradov S A and Charpak S 2011 Simultaneous two-photon imaging of oxygen and blood flow in deep cerebral vessels *Nat. Med.* **17** 893

[94] Sud D and Mycek M-A 2009 Calibration and validation of an optical sensor for intracellular oxygen measurements *J. Biomed. Opt.* **14** 020506

[95] Jahn K, Buschmann V and Hille C 2015 Simultaneous fluorescence and phosphorescence lifetime imaging microscopy in living cells *Sci. Rep.* **5** 14334

[96] Kalinina S, Schaefer P, Breymayer J, Bisinger D, Chakrabortty S and Rueck A 2018 Oxygen sensing PLIM together with FLIM of intrinsic cellular fluorophores for metabolic mapping *Proc. SPIE* **10497** 104970F

[97] Sud D, Mehta G, Mehta K, Linderman J, Takayama S and Mycek M-A 2006 Optical imaging in microfluidic bioreactors enables oxygen monitoring for continuous cell culture *J. Biomed. Opt.* **11** 050504

[98] Howard S S, Straub A, Horton N G, Kobat D and Xu C 2013 Frequency-multiplexed *in vivo* multiphoton phosphorescence lifetime microscopy *Nat. Photonics* **7** 33

Chapter 10

Future outlook and conclusions

This book has covered a number of topics that form the state-of-the-art in VLC systems, ranging from a general introduction in chapter 1 through to using VLC in a medical setting in a brand new way. This book should go some way towards demonstrating the versatility of VLC systems and the applications that it can support.

These include, but are not limited to, high speed data broadcasting, localisation and positioning, intelligent transport systems and the internet-of-things, which are covered in chapters 4, 7 and 8, respectively. The range of versatile skills that are required to support such a large range of applications are reflected in the expected market value of VLC in the UK alone, which is estimated at over £6 billion today, in 2020.

The book has covered the foundations of VLC including the fundamental semiconductor technologies that underpin the operation of the LEDs and PDs that it relies on to operate correctly. Without these technologies, the domain would not exist and therefore, a solid foundation must be obtained by anyone who wishes to enter the VLC field. Since the majority of my 10 years researching within the VLC domain have been focused on organic devices such as PLEDs and OPDs [1], organic semiconductors were also included in chapter 2, which will hopefully inspire researchers in the future to look at these (both figuratively and literally) amazingly flexible technologies and create new applications that drive the domain forward.

Chapter 3 then looked at high speed circuit design and discussed several of the key techniques that have been presented in the literature by Le Minh [2], Odedeyi [3] and Kassem [4], along with their collaborators. There have been numerous reports in the literature on circuit designs to achieve high speed links; however, these are the three most promising in my opinion. In particular, the negative impedance conversion approach that extends the bandwidth of LEDs with no resistive loading, and therefore no power penalty [4], which could be extremely important in the future considering the high speeds already achieved, notably from Chi's research group in

doi:10.1088/978-0-7503-1680-4ch10

Fudan University, China, who extensively rely on resistive-equalisation techniques and yet still set numerous world record transmission rates [5]. Chapter 3 also examined a highly cited paper on channel modelling by Chvojka and his collaborators, where he modelled the movement of people through some common room scenarios and settings, accurately modelling their influence on supporting VLC links.

High speed modulation is currently one of the hottest topics within the research domain, and modulation formats are becoming more 'exotic' in order to squeeze every available fraction of a bit/second/Hz. Therefore, rather than focus on traditional modulation formats, chapter 4 focused on new and upcoming modulation formats and some applications that were reported in the literature [6], and supplemented it with a discussion of machine learning-based equalisers. Machine learning is an emerging topic that is gaining more and more traction due to its ability to solve the majority of problems with ease, albeit at high computational complexity.

As VLC is normally promoted as a dual function, dual advantage technology providing lighting in combination with communication systems, chapter 5 focused on dimming whilst sustaining transmission. The relationships between perceived and measured optical power were described and colour balancing issues were discussed, focusing on an excellent report by Popoola [7] that also looked at health implications of light flickering and colour balancing.

One of the biggest criticisms of VLC, in my opinion, is the lack of a serious contender for the uplink technology, if VLC is to be used as an access network in a 5G and/or beyond setting. There is a strong argument to use any one of three technologies, either Wi-Fi, IR or VLC, and I personally believe that a hybrid approach that makes use of the advantages of each candidate technology is the best way to move forward. Therefore, the LiRa report in [8] deserves particular attention due to the development of a hybrid MAC that establishes the first stepping stones towards this in a multi-user environment with a real stack deployment.

The future looks bright for the VLC domain but all of the above applications still have significant challenges remaining. I personally believe that the key applications are in the final three topics discussed here in these closing statements and I am excited to see where the future brings the technology.

References

[1] Minotto A, Haigh P A, Łukasiewicz Ł G, Lunedei E, Gryko D T, Darwazeh I and Cacialli F 2020 Visible light communication with efficient far-red/near-infrared polymer light-emitting diodes *Light Sci. Appl.* **9** 1–11

[2] Le Minh H, O'Brien D, Faulkner G, Zeng L, Lee K, Jung D and Oh Y 2008 High-speed visible light communications using multiple-resonant equalization *IEEE Photonics Technol. Lett.* **20** 1243–5

[3] Odedeyi T, Haigh P A and Darwazeh I 2018 Transmission line synthesis approach to extending the bandwidth of LEDs for visible light communication *2018 11th Int. Symp. on Communication Systems, Networks Digital Signal Processing (CSNDSP)* pp 1–4

[4] Kassem A and Darwazeh I 2019 Exploiting negative impedance converters to extend the bandwidth of LEDs for visible light communication *2019 26th IEEE Int. Conf. on Electronics, Circuits and Systems (ICECS)* pp 302–5

[5] Zhu X, Wang F, Shi M, Chi N, Liu J and Jiang F 2018 10.72 Gb/s visible light communication system based on single packaged RGBYC LED utilizing QAM-DMT modulation with hardware pre-equalization *Optical Fiber Communication Conf.* (Optical Society of America) p M3K-3

[6] Stepniak G 2018 Staggered CAP—a new spectrally efficient modulation format for optical communications *IEEE Photonics Technol. Lett.* **30** 367–70

[7] Popoola W O 2016 Impact of VLC on light emission quality of white LEDs *J. Lightwave Technol.* **34** 2526–32

[8] Naribole S, Chen S, Heng E and Knightly E 2017 LiRa: a WLAN architecture for visible light communication with a Wi-Fi uplink *2017 14th Annual IEEE Int. Conf. on Sensing, Communication, and Networking (SECON)* (IEEE) pp 1–9

Lightning Source UK Ltd.
Milton Keynes UK
UKHW050713270421
382700UK00003B/160

9 780750 316781